A Colour Atlas

Carbonate Sediments and Rocks Under the Microscope

A. E. Adams
Formerly Senior Lecturer in Geology
University of Manchester, England

W. S. MacKenzie
Late Emeritus Professor of Petrology,
University of Manchester, England

CRC Press
Taylor & Francis Group
Boca Raton London New York

CRC Press is an imprint of the
Taylor & Francis Group, an **informa** business

CRC Press
Taylor & Francis Group
6000 Broken Sound Parkway NW, Suite 300
Boca Raton, FL 33487-2742

First issued in paperback 2018

© 1998 by Taylor & Francis Group, LLC
CRC Press is an imprint of Taylor & Francis Group, an Informa business

No claim to original U.S. Government works

ISBN 13: 978-1-138-43017-4 (hbk)
ISBN 13: 978-1-874545-84-2 (pbk)

This book contains information obtained from authentic and highly regarded sources. Reasonable efforts have been made to publish reliable data and information, but the author and publisher cannot assume responsibility for the validity of all materials or the consequences of their use. The authors and publishers have attempted to trace the copyright holders of all material reproduced in this publication and apologize to copyright holders if permission to publish in this form has not been obtained. If any copyright material has not been acknowledged please write and let us know so we may rectify in any future reprint.

Except as permitted under U.S. Copyright Law, no part of this book may be reprinted, reproduced, transmitted, or utilized in any form by any electronic, mechanical, or other means, now known or hereafter invented, including photocopying, microfilming, and recording, or in any information storage or retrieval system, without written permission from the publishers.

For permission to photocopy or use material electronically from this work, please access www.copyright.com (http://www.copyright.com/) or contact the Copyright Clearance Center, Inc. (CCC), 222 Rosewood Drive, Danvers, MA 01923, 978-750-8400. CCC is a not-for-profit organization that provides licenses and registration for a variety of users. For organizations that have been granted a photocopy license by the CCC, a separate system of payment has been arranged.

Trademark Notice: Product or corporate names may be trademarks or registered trademarks, and are used only for identification and explanation without intent to infringe.

Visit the Taylor & Francis Web site at
http://www.taylorandfrancis.com

and the CRC Press Web site at
http://www.crcpress.com

CONTENTS

Preface	4
Acknowledgements	5
Introduction	6
Coated Grains	9
Peloids, Aggregate Grains, Intraclasts and Lithoclasts	23
Bioclasts	32
Diagenesis	101
Porosity	156
Limestone Classification	164
Cathodoluminescence	168
Bibliography	176
Index	177

PREFACE

Examination of thin sections under the microscope is a key part of any study of carbonate sediments, as a companion to field or core logging, and as a necessary precursor to geochemical analysis. This book is designed as a laboratory manual to keep beside the microscope as an aid to identifying grain types and textures in carbonates. For the newcomer to the subject, carbonates can display a bewildering variety of grains, compared to sandstones, for example, and we hope this book will help to give confidence to those initial observations. By illustrating more than one example of common grains and textures, we hope that the more experienced practitioner will also find assistance in identifying the unfamiliar. However, such is the diversity of carbonate sediments, that it is impossible to be completely comprehensive and if we have omitted your favourite bioclast, then sorry! Throughout we have tried to show good, but typical rather than exceptional, examples of each feature. It has not been our intention to supply much interpretation except where this is necessary to explain the origin of features illustrated.

Two comments that we have received about previous atlases (Adams *et al.*, 1984; MacKenzie & Adams, 1994) are worth mentioning here. Firstly, it is possible to claim that some photographs are over- or underexposed. Photography of carbonate sediments can be difficult, especially where there are micritic grains, which are almost opaque, set in a coarsely crystalline, clear sparite cement. The ex-

posure has to be adjusted, such that, if the micritic grains are the subject of the picture, the cement may well appear overexposed, and if the cement is the subject, the grains will be underexposed and appear almost black. When using the microscope, the eye makes adjustments depending on what you are looking at, and in any case it is easy, and often necessary with carbonates, to vary the light intensity.

Secondly, we have been asked why we have not supplied a full petrographic description of a rock. We do not believe that this can be usefully done from a photograph, typically showing a field of view a few millimetres across. Carbonate rocks can vary such that no one field of view is representative of the whole rock This is particularly true of coarser packstones and grainstones with a mixture of grain types. Those wanting a format for a full petrographic description are directed to Flügel (1982) and Harwood (1988).

We have avoided using abbreviations in the text, but in each plate caption we have used the abbreviation 'PPL' for plane-polarised light and 'XPL' for pictures taken with polars crossed.

Finally, we hope that readers whatever their level of interest in carbonate sedimentology, will, by browsing through this atlas, be able to appreciate just a little of the wonder of the geological world as revealed under the microscope, and share our excitement at the beauty and variety of natural textures seen in carbonates.

ACKNOWLEDGEMENTS

No book such as this can be completed without the loan of material from generous colleagues. While much of the material illustrated here comes from the collections of the Department of Earth Sciences, University of Manchester, we are very grateful to the following for collecting material from which we could have thin sections made, or for loaning us their own thin sections: Waleed Abdulghani, Alham Al-Langawi, Pat Cossey, Alistair Gray, Pete Gutteridge, Andrew Horbury, Dave Hunt, Kieron Jenkins, Rhodri Johns, Joe Macquaker, Neil Pickard, Tony Ramsay, Kevin Schofield and Fiona White.

We would particularly like to thank Dave Hunt for reading the whole manuscript and for making many useful comments on the text and pictures, and Tony Ramsay who checked the section on Foraminifera. While their advice has much improved the layout and content of the book, any errors and mis-interpretations are entirely the authors' responsibility.

INTRODUCTION

This book is arranged with depositional features first, followed by diagenesis, although the final appearance of a carbonate rock is often as much the product of the secondary, diagenetic processes it has suffered, as it is of primary depositional processes. For this reason, diagenetic features visible in the photographs of depositional features are commented on in the text as well as *vice-versa*. In most cases we have cross-referenced to the definition of terms used, and we have provided an index for quick reference to the main descriptions and illustrations of technical terms.

Carbonate rocks can be regarded as having six main components:

- The grains, comprising discrete, organized aggregates of calcium carbonate, sometimes also known as the allochemical components or allochems. These comprise the coated grains, peloids, intraclasts and bioclasts etc., and are considered in the first part of the book.
- Carbonate mud sediment occurring as matrix between the grains. This is the clay and silt-sized particles deposited along with the grains or infiltrated into the sediment in the depositional environment. It has many origins – some may be a precipitate from supersaturated sea water and some undoubtedly forms from the comminution of other grains. The very finest material (clay-sized, <5 m), also known as micrite (short for microcrystalline calcite) is probably either a precipitate or has formed from the disintegration of encrustations around organisms such as green algae.
- Terrigenous components, consisting of detrital quartz, clay or other non-carbonate minerals. These are not considered separately in this book.
- Sparry calcite, sparite or spar refer to the larger calcium carbonate crystals which are a pore-filling cement and thus part of the secondary diagenetic story.
- Replacive crystals of dolomite, evaporite minerals or other non-carbonates. These are also diagenetic.
- Pore-space, referring to any spaces in the sediment, filled with air, water or hydrocarbons.

Classification of limestones involves estimating the proportions of these components. We have placed classification near to the end of this book,

because in our discussion of the problems of limestone classification we have made reference to diagenetic features mentioned in the later parts of the text. In our plate descriptions, however, we have used names based on the Dunham classification according to depositional texture (p.164).

The mineralogy of carbonate sediments and rocks is more complex than it might first appear. Recent marine carbonates consist of aragonite and calcite (with variable Mg content) and small amounts of dolomite. Aragonite is metastable in sedimentary conditions and ancient carbonates consist of calcite (with little Mg) and dolomite. Both calcite and dolomite in ancient sediments may contain some iron when they are said to be 'ferroan'. An important part of any study of diagenesis is understanding how and when aragonite is dissolved or altered to calcite, the nature of the replacement of calcium carbonate by dolomite and the introduction of ferroan minerals.

Although modern shallow marine carbonate sediments contain a lot of aragonite, this was not the case in the past. Today many of the common organisms, such as the corals, many molluscs and green algae have hard parts of aragonite. At many times in the past organisms with calcitic hard parts dominated. Furthermore it seems that the mineralogy of precipitated shallow marine carbonate may have varied throughout the Phanerozoic as a result of changes in seawater composition related to global sealevels and the rate of global processes such as sea floor spreading. It seems that for much of the Phanerozoic, calcite was the preferred precipitate, the exceptions being between the Late Carboniferous and Late Triassic and from the Early Tertiary to the present day when aragonite was dominant. The evidence for this comes, in part, from the study of ooids, grains in which chemical precipitation is thought to play a major part The reader is directed to Tucker & Wright (1994) for a discussion of the nature and causes of the temporal variation in carbonate mineralogy.

Staining

Because of the similarity in optical properties of calcite and dolomite, about half the thin sections illustrated in this book have been etched and stained, using two stains dissolved in weak hydrochloric acid, according to the method described by Adams *et al.* (1984) adapted from Dickson (1965). The two stains used are Alizarin Red S, to help distinguish calcite from dolomite, and potassium ferricyanide which distinguishes carbonate minerals containing Fe^{2+} (ferroan minerals) from those with little or no iron (non-ferroan minerals). The results of the staining procedure are shown in **1**. The following points should be noted:

- The etching introduces a differential relief. This is particularly marked in sediments which have mixtures of calcite and dolomite. The calcite reacts with the dilute acid and is reduced in relief in comparison to dolomite which does not significantly react with weak acid. However, some differential relief may be introduced in limestones, because the calcite crystals will be in different orientations. Those orientations which have most cleavages intersecting the surface of the section react more rapidly with the acid and are thus reduced in relief.
- Description of colour is a very subjective matter. Years of teaching students have shown us that the turquoise colours of ferroan dolomite are called green by many people, and the blues of ferroan calcite, turquoise. It does not matter what you call the colours so long as you become familiar with the shades developed by each mineral species.
- The intensity of the colours developed depends on the degree of reaction with the acid. Thus a ferroan dolomite will be a much paler colour than a ferroan calcite with the same amount of iron

substitution, because the dolomite has minimal reaction with the acid. Similarly, fine-grained fabrics react more rapidly with acid because the acid attacks along crystal boundaries and thus carbonate muds show deep colours compared with coarser fabrics of the same composition in the same rock. Indeed, some calcite sparites may appear unstained and are interpreted as dolomites by the beginner, but careful inspection usually reveals pink spots across the crystal or a red stain along crystal boundaries. Those sparry crystals with the least stain, those which react least with the acid, are large crystals in an orientation where fewest cleavages intersect the surface of the thin section. This orientation is one where the viewer is looking down a line close to the optic axis of calcite. In this orientation, as well as being lightly stained or unstained, the crystals show much lower order interference colours than normal.
- The blue colour developed with a ferroan calcite is a much more intense colour than the red, so that in iron-rich fabrics the blue may mask the red completely. Furthermore, because of its intensity, the blue colour of ferroan calcite is obvious with coarse-grained fabrics, while the pink of coarse-grained calcite may not be (see above).

Many examples of stained thin sections are described in this book so just one example is illustrated here (**2**). This limestone comprises bundles of worm tubes (but see discussion on p.96). The tubes themselves are made of fairly fine-grained non-

Mineral	Effect of etching	Stain colour with Alizarin Red S	Stain colour with potassium ferricyanide	Combined result
Calcite (non-ferroan)	Considerable (relief reduced)	Pink to red-brown	None	Pink to red-brown
Calcite (ferroan)	Considerable (relief reduced)	Pink to red-brown	Pale to deep blue depending on iron content	Mauve to blue
Dolomite (non-ferroan)	Negligible (relief maintained)	None	None	Unstained
Dolomite (ferroan)	Negligible (relief maintained)	None	Very pale blue	Very pale blue (appears turquoise or greenish in thin section)

1 Summary table to show results of etching and staining carbonate minerals

Carbonate Sediments and Rocks Under the Microscope

ferroan calcite and are thus stained red. There is some micrite sediment in the lower part of the section and, although it is the same mineralogy as the worm tubes, it has a much more intense stain (brown) because of its finer grain size. Both ferroan calcite and ferroan dolomite are present in this rock. The ferroan calcite occurs as a sparite cement and is best seen filling the three tubes near the centre of the picture and where it is a typical mauvey-blue colour. It also occurs as thin veins cutting the tubes, for example near the top of the picture. The ferroan dolomite is a much paler greeny-blue (turquoise) and occurs within and between tubes, especially in the upper third of the photograph.

The following photographs also illustrate the stain colours particularly well:

18 Coarse-grained ferroan and non-ferroan calcites.
65 Variation of stain intensity according to crystal size and orientation.
91 Mauve-stained slightly ferroan calcite.
248, 252, 256, 268 Dolomite and calcite.
272 Ferroan dolomite.
275 Ferroan dolomite, ferroan and non-ferroan calcite.
282 Calcites with varying iron content.

Two other staining techniques have been employed for material illustrated in this book. A very important technique, particularly in the petroleum industry, is the impregnation of porous rocks with a blue-dye-stained resin. The impregnation helps to keep porous and sometimes rather friable rocks from breaking up during section-making and the blue dye makes recognition and evaluation of the porosity straightforward. The blue colour is similar to that of some potassium ferricyanide stained ferroan calcites and to avoid possible confusion, sections impregnated with blue-dye-stained resin shown here are otherwise unstained. Examples of thin sections prepared in this manner are illustrated in **57, 87, 193, 196, 210, 261, 267**, and in the porosity section (p.156).

Stains can also be employed to detect aragonite and to distinguish high-Mg from low-Mg calcite although their use has largely been superseded by analytical techniques. Feigl's Solution is an alkaline solution of silver and manganese salts which when applied to sections gives a spotty black colour with aragonite while calcite remains untouched. The results are illustrated in **46**.

2 Stained thin section, Lower Carboniferous, Northumbria, England, PPL, × 50.

COATED GRAINS

Amongst the non-skeletal carbonate grains, the variety of coated grains has perhaps received most coverage in the literature. Coated grains comprise a more or less well-defined nucleus, surrounded by a coating of calcareous material, usually fairly fine-grained, called the cortex. In many coated grains the cortex is, at least in part, laminated. Many different names and definitions have been used for coated grains (see, for example, Peryt, 1983) with classifications based on size, shape, regularity of concentric laminations, presence of obvious biogenic structures and, often, an interpretation of their origin. There is also, as exists elsewhere in carbonate sedimentology, a discrepancy between what is known from Recent environments and what is found in the geological record.

Since coated grains of similar appearance can form in different environments, we favour a simple descriptive classification following the usage of Tucker & Wright (1990), rather than one which relies on an interpretation of their origin. Where appropriate, the likely origin of the illustrated grains is discussed in the plate captions.

Ooids and pisoids are spherical or ellipsoidal coated grains with a nucleus surrounded by a cortex of which at least the outer part is smoothly concentrically laminated. Obvious biogenic structures should be at most a minor component of the cortex. Ooids are <2 mm in diameter, pisoids >2 mm.

In older literature the terms oolith and pisolith are used, but have largely been replaced by ooid and pisoid. However, the terms oolitic for a sediment containing ooids, and oolite for a sediment dominated by ooids, are still in use. To increase the confusion the adjective ooidal also gets an occasional airing! There is also some variation in spelling (and in pronunciation), with Bathurst in his seminal textbook (Bathurst, 1975) using oöid and oölite throughout.

Most ooids and pisoids are calcareous, but there are many instances of grains fulfilling the definitions of ooids and pisoids with non-calcareous cortices. Best known are the grains from oolitic ironstones in which the cortices comprise various iron minerals. Only calcareous ooids and pisoids are discussed here. Illustrations of oolitic ironstones can be found in Adams et al. (1984).

Oncoids are coated grains in which the calcareous cortex is less smoothly laminated than in the case of ooids and pisoids, with irregular laminae overlapping and often not entirely concentric. Oncoids are often irregular in shape, may have a poorly defined nucleus and may contain biogenic structures. Oncoids are >2 mm in diameter; grains of this type <2 mm in diameter are known as micro-oncoids.

Oncolith is a synonym for oncoid and the terms oncoidal, oncolitic and oncolite are all in use.

Ooids

Ooids demonstrate a variety of cortical structures and mineralogies which depend on their age, mode of formation and diagenesis. Recent marine ooids are mostly made of aragonite, but with a variety of microfabrics. Within the concentric laminae of the ooids, aragonite needles or rods may be tangentially, randomly or radially arranged. Individual ooids may show mixtures of laminae with different microfabrics. Classic ooids from carbonate sand shoals of the Bahamas and many ooids from the Arabian Gulf have dominantly tightly-packed tangential fabrics. These seem to characterise the highest-energy ooid-forming areas where perhaps grain-to-grain collisions impact the fabric or break off any crystals in non-tangential alignments.

3–5 show a grain mount of Recent ooids from the Arabian Gulf. Two ooids have detrital quartz nuclei, recognisable by the low relief in the photograph taken with plane-polarised light (3) and the first-order grey interference colours visible under crossed polars (4). The ooid cortices show the typical brownish translucent appearance of Recent marine ooids. The colour may result from included organic matter which probably plays an important part in their formation. Unresolved questions regarding ooid genesis include the relative importance of organic (microbial) processes and inorganic precipitation. Because the aragonite rods which comprise the tangentially orientated layers are typically of the order of a micron in length, individual crystals cannot be seen with a light microscope. However, the compact aligned fabric is responsible for the translucent appearance. Randomly orientated or equant micron-sized crystals appear almost opaque in sections of normal thickness, as for example in the nucleus of the ooid seen on the lower edge of 3. Darker areas within the ooid cortices are probably also areas of random or equant aragonite, perhaps produced through micritisation by endolithic micro-organisms (p.101).

Even though the individual crystals cannot be seen, the preferred orientation can be demonstrated. In 4, taken with polars crossed, ooids show an extinction cross, sometimes referred to as a pseudo-uniaxial cross, because of its resemblance to the interference figure seen when looking down the optic axis of a uniaxial mineral with convergent light.

Aragonite is orthorhombic and has straight extinction, so that rod-shaped crystals go into extinction when orientated N–S or E–W. An ooid with a radial structure will therefore show a well-developed extinction cross, at least when the plane of the section passes through or near to the centre of the grain. In tangential ooids, the aragonite crystals are randomly orientated on the tangential surface. Although crystals will therefore be cut in many different sections, they are all in orientations which are at extinction or have low double refraction, and thus an extinction cross is seen. The presence of a dominantly radial or tangential orientation to the constituent crystals is therefore responsible for the extinction cross. Despite the high-order interference colours usually seen in sections through carbonate minerals of normal thickness, modern ooids typically show low-order interference colours resulting from the small size of the crystals, which are much thinner than the section as a whole. In 4 the ooid aragonite is showing first-order orange interference colours.

In 5 a length-slow sensitive tint plate has been inserted NW–SE. Careful inspection of the interference colours (second-order yellow in SE and NW quadrants, first-order grey in NE and SW quadrants) shows that the length-fast aragonite is tangentially orientated. Calcite is also length fast in elongated crystals so there is no need to determine mineralogy before orientation.

Coated Grains

3–5 Grain mount thin section, Quaternary, Arabian Gulf, ×30, **3** PPL, **4** XPL, **5** XPL with sensitive tint plate.

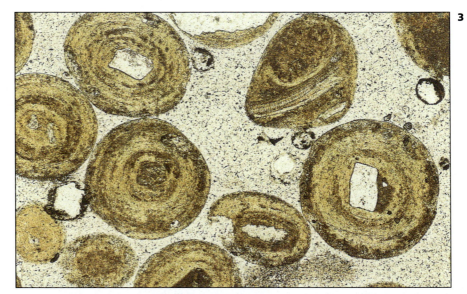

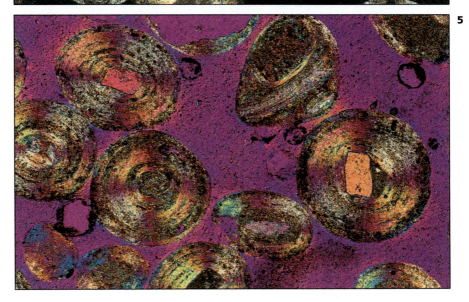

6 shows aragonite ooids cemented in a Recent limestone from the Bahamas. Despite the aragonite having been exposed to meteoric water, there is no visible sign of solution or alteration as yet (cf. **16**). The smooth concentric lamination is more clearly seen in these ooids than in those from the Arabian Gulf (**3–5**). The nuclei are entirely calcareous and include almost opaque micritic grains (peloids) of rather large size compared to the thickness of the cortex.

Ooids from the Great Salt Lake in Utah show a variety of aragonitic microfabrics apparently dependent on the degree of organic influence and agitation during their formation. The large grain below and to the right of the centre of **7** shows well-developed radial and concentric structures and resembles calcitic counterparts in many ancient limestones (e.g. **8**). Other grains in the photograph show much coarser radial fabrics with an irregular outline and an absence of well-defined concentric laminae. Grains lacking concentric laminae are not strictly ooids, but can be called spherulites if a radial structure is present. The larger, radially arranged, elongate aragonite crystals with a subcircular cross-section have been called 'clubs'.

The fabrics shown by ancient ooids depend on their original structure and mineralogy and on their subsequent diagenesis. The best-preserved ooids in the geological record are calcite with a well-developed radial structure. By comparison with Recent ooids from the Bahamas, such ooids were once thought to have altered from original tangential aragonite ooids, and elaborate mechanisms were devised to account for this change. However, the consensus now is that these were primary calcite radial ooids. The preponderance of calcitic ooids at certain times in the past has led geologists to suggest that sea-water chemistry has changed, at times promoting calcite precipitation and at other times aragonite.

At least some radial ooids appear to have formed in somewhat lower energy environments than those with a tangential structure and they are often less smoothly laminated. **8** shows ooids with a well-developed radial structure, the largest of which is rather irregularly laminated. The ooids also contain layers and irregular areas of dark, probably equant, micritic calcite.

9 shows two broken ooids, one of which has acted as the nucleus for further ooid growth. Such breakage is not uncommon with radial ooids, but rare in ooids with other structures. Perhaps the radial structure with rather coarser crystals is mechanically weaker than the tightly bound, fine tangential structure.

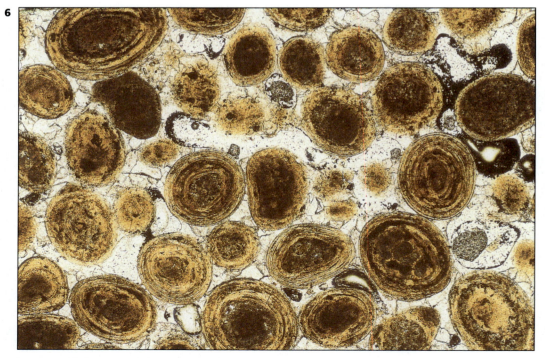

6 Unstained thin section, Quaternary, Bahamas, PPL, × 50.

Coated Grains

7 Grain mount thin section, Quaternary, Great Salt Lake, Utah, USA, PPL, × 25.

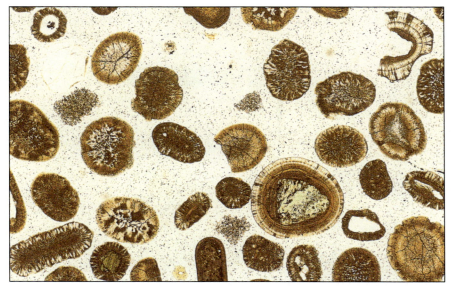

8 Unstained thin section, Upper Jurassic, Provence, France, PPL, × 30.

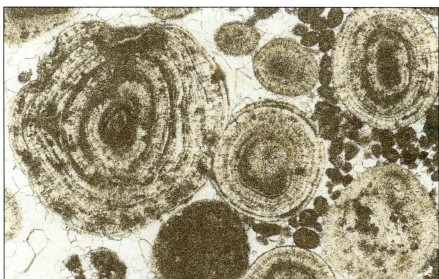

9 Unstained thin section, Upper Jurassic, Provence, France, PPL, × 30.

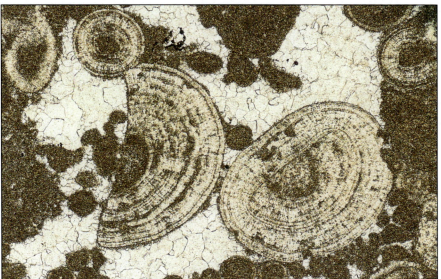

Carbonate Sediments and Rocks Under the Microscope

Radial ooids are rarely as well preserved as those in **8** and **9**. The ooid preservation in **10** is more usual, with radial layers having been partly altered to equant micritic calcite, probably by microbial micritisation.

Ooids with only one or two concentric layers are known as superficial ooids. Where the ooid nucleus is irregular in shape, early laminae are not concentric, but tend to round off the shape of the grain such that sphericity increases with growth. The ooids in **11** and **12** are superficial ooids with a radial structure in the cortex. Many have detrital quartz nuclei and show how the initially irregular quartz grain surface has become rounded during formation of the early ooid laminae.

Tangential calcite fabrics appear to be rare or absent in the geological record of ooids. Micritic ooids retaining some concentric structure, but with equant crystals of calcite are common, although it is difficult to know whether this is a primary or secondary fabric and, if secondary, what the original fabric and mineralogy were. **13** shows ooids of this type. The equant micritic crystals result in an almost opaque grain, in contrast to the translucent nature of ooids with elongate crystals in a preferred orientation (**3, 6, 8**).

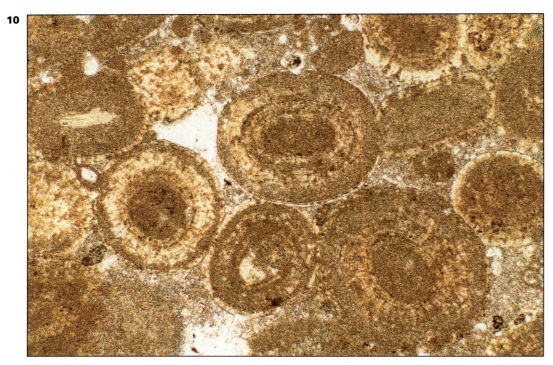

10 Stained thin section, Middle Jurassic, Mallorca, Spain, PPL, × 50.

Coated Grains

11, **12** Stained thin section, Upper Jurassic, Western High Atlas, Morocco, × 30, **11** PPL, **12** XPL.

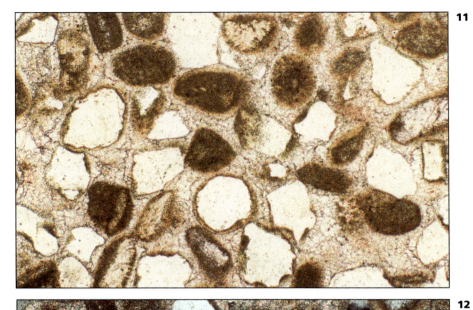

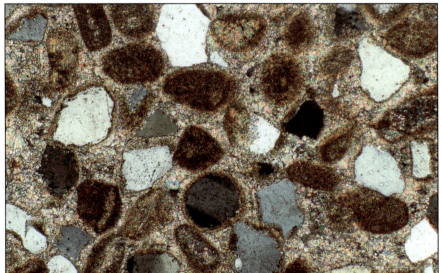

13 Stained thin section, Lower Carboniferous, South Wales, PPL, × 50.

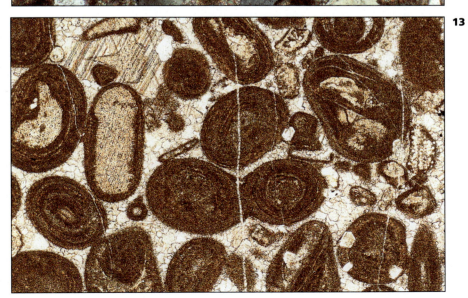

Carbonate Sediments and Rocks Under the Microscope

Aragonite ooids are susceptible to solution or alteration to neomorphic spar (p.128), particularly if exposed to meteoric water. **14** and **15** show pores of roughly circular section which, based on their size and sphericity, are interpreted as having been formed as a result of the solution of ooids. The pores appear speckled in the photograph taken with plane-polarised light (**14**); the speckles are bits of the grinding powder which have stuck to the mounting medium. The pores are evident in the view with crossed polars (**15**) and the sensitive tint plate inserted, showing violet. This is an 'inside out' rock, the original grains having been totally dissolved and the original water-filled pores having been filled with cement. Clearly the cementation must have occurred prior to solution of the ooids in order to maintain the texture of the rock. This sediment is said to have high oomouldic porosity, although its permeability will be low if the pores are not interconnected.

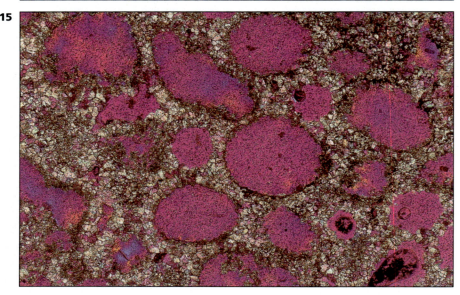

14, **15** Unstained thin section, Miocene, Mallorca, Spain, × 25, **14** PPL, **15** XPL with sensitive tint plate.

Coated Grains

16 and **17** show ooids that have suffered a combination of solution and neomorphic alteration. The smaller grains in the right centre and upper part of the photograph have completely dissolved, leaving moulds of nearly circular section, whereas the larger ooid with the quartz nucleus is mostly altered to neomorphic calcite (see p.128 for definition and discussion of neomorphism). It retains inclusions which demonstrate the original concentric structure, but the cortex has been altered to comprise large irregular calcite crystals which cut across a number of the original laminae. This is most clearly seen in the crossed polars view with the sensitive tint plate inserted (**17**). The centre of the ooid in the lower right corner of the photograph appears to have dissolved, but the outer part of the cortex is neomorphic spar.

16, **17** Unstained thin section, Quaternary, Florida, USA, × 40, **16** PPL, **17** XPL with sensitive tint plate.

Carbonate Sediments and Rocks Under the Microscope

In **15** the ooid nuclei have disappeared and are presumed to have been aragonite as well as the cortex. In contrast, **18** shows casts of ooids that had calcite echinoderm plate nuclei of rather large size compared to the cortex. The cortices, presumably of aragonite originally, dissolved to leave the calcite nuclei. These would have fallen to the bottoms of the oomoulds, something not clearly shown by the picture. The nuclei are to the sides, but show no consistent pattern. This may be because the rock was not cut normal to bedding. The moulds were later infilled by a ferroan calcite cement to become casts, and are clearly picked out by the blue stain colour.

Dolomitic sediments often contain replaced ooids which can sometimes be identified if the replacement has mimicked the original rock texture to some degree (p.136). **19** shows a completely dolomitised sediment which retains 'ghosts' of rounded grains. When dolomite mimics calcite fabrics, crystal size variation normally reflects crystal size in the original limestone. In this case, the original spherical grains are now coarsely crystalline dolomite, suggesting that it may have been a sediment containing oomoulds or oocasts that was replaced by dolomite.

18 Stained acetate peel, Lower Carboniferous, South Wales, PPL, × 25.

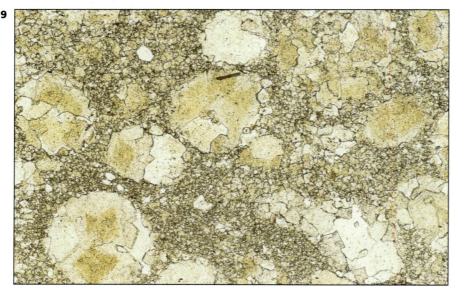

19 Unstained thin section, Middle Jurassic, Western High Atlas, Morocco, PPL, × 25.

Pisoids

Smoothly laminated grains greater than 2 mm in diameter are much less common than ooids and form in more specialized environments. Normal marine ooids probably do not grow to large sizes because carbonate layers are accreted while the grain is held in suspension and grains larger than 2 mm cannot be held in suspension for long enough to be coated. Many pisoids grow, at least in part, within the sediment and there is no such restriction in size during growth. **20** and **21** show small pisoids, each about 3 mm in diameter. The sediment in **20** is completely dolomitised and the smooth concentric laminations are picked out by variations in dolomite crystal size. It is not possible to interpret the origin of this grain from the photograph alone.

Some pisoids form partially or completely in the vadose zone, above the water-table. The pisoid in **21** shows evidence of having been cemented in the vadose zone, with early cement present on the undersurface of the grain and absent from the upper surface, but does not itself show clear evidence of preferential downwards growth, a characteristic of vadose pisoids (vadose textures are described in more detail and illustrated in **197—202**). Another characteristic of pisoids that have grown *in situ*, at least during the later stages of growth, is that individual grains may coalesce to form composite grains. Repeated desiccation of vadose pisoids can lead to radial fracture and grain breakage. Fragments of pisoids can be seen along the left-hand edge of **21**.

20 Unstained thin section, Permian, Yorkshire, England, PPL, × 22.

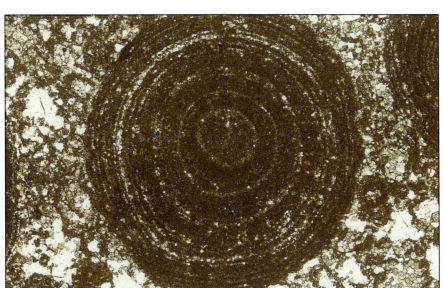

21 Stained thin section, Lower Jurassic, Greece, PPL, × 16.

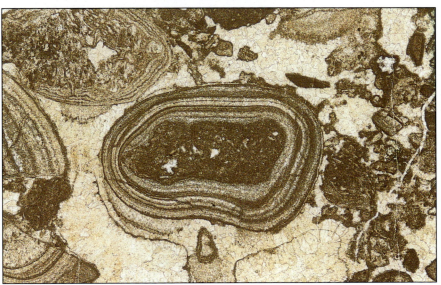

Oncoids

Although the definitions of coated grains adopted here imply no particular mode of formation, most oncoids are undoubtedly biogenic in origin. Their laminations are less regular than those of ooids and pisoids and are often more obvious in hand specimen than under the microscope. This is because, rather than being partly defined by layers of different microfabric, many oncoid laminae are defined by differences in colour which may reflect differences in amounts of organic matter or other impurities. Many dominantly micritic oncoids were probably formed by cyanobacteria and were formerly known as algal nodules. **22** shows an oncoid from a marine Jurassic limestone. The nucleus is a thin piece of brachiopod shell, the cortex is indistinctly laminated and, in contrast to ooids, there is no tendency to increased sphericity during growth. There are also other organisms within the cortex, including bryozoans along the lower side. It is not unusual to find attached or encrusting organisms within the cortices of oncoids. Not all oncoids are marine: **23** is from a freshwater limestone. In this case the nucleus is rather indistinct, but the concentric laminations and the rather irregular outer surface of the grain are clear.

24 shows part of an oncoid that was about 1 cm in diameter. The centre was at some stage hollow and has been filled with cement, initially of pink-stained non-ferroan calcite and later of blue-stained ferroan calcite (lower left). The micritic coating is interrupted by a layer of sparry calcite which is probably a precipitate that formed on the surface of the grain during growth. The micrite of the coating is typically rather blotchy with little irregular patches of spar. Some of these areas have a filamentous appearance (for example, just to the right of centre) and these may be casts of the microbial organisms involved in the construction of the grain.

Other coated grains

Despite the simple definitions, some coated grains may be difficult to classify. In this section we illustrate some of these 'problematic' grains and discuss their nomenclature and origin. The grain in **25** is 1.5 mm in diameter and is smoothly concentrically laminated. It has a well-defined nucleus consisting of a gastropod with a quartz grain lodged in its aperture. However, it has an unusually high proportion of attached or encrusting bioclasts within the cortex for an ooid, and is too smoothly laminated to be termed a micro-oncoid. In such cases, careful description of grains is more important than worrying about what to call them. Other examples of coated grains may be regarded as intraclasts and are illustrated in **35** and **37**.

22 Stained thin section, Middle Jurassic, Cotswolds, England, PPL, × 10.

Coated Grains

23 Unstained thin section, Upper Jurassic, Palencia, Spain, PPL, × 20.

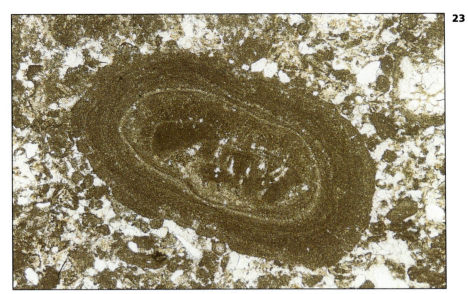

24 Stained thin section, Lower Carboniferous, South Wales, PPL, × 20.

25 Unstained thin section, Jurassic, Greece, PPL, × 40.

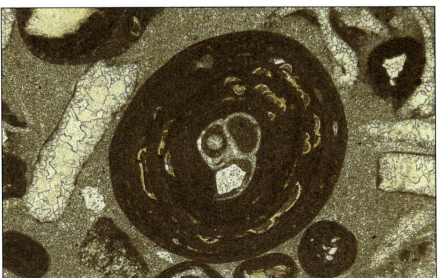

Ooids and pisoids have frequently been described from paleosols, where they form largely through microbial processes. **26** and **27** show coated grains from fossil soils. Those in **26** are small, but are closer to the definition of micro-oncoids than that of ooids. Detrital quartz is an important component of this sediment. The complex aggregate grains in **27** show evidence of *in situ* growth, at least during the later stages of their formation, with smaller grains enveloped in the coatings of larger ones and bridging sediment between grains or groups of grains. Other features of fossil soils are illustrated in **188–191**.

Rhodoids are often considered to be coated grains. They are nodules dominated by coralline algae, but although they may show a concentric structure and have a nucleus, they are not non-skeletal particles and are considered here with the bioclasts (p.89).

26 Stained thin section, Quaternary, Morocco, PPL, × 40.

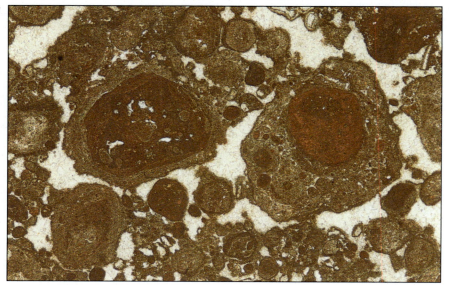

27 Stained thin section, Lower Jurassic, Greece, PPL, × 30.

PELOIDS, AGGREGATE GRAINS, INTRACLASTS AND LITHOCLASTS

In addition to coated grains there are many other types of non-skeletal particle described from limestones. The nomenclature used here broadly follows that of Tucker & Wright (1990).

A peloid is a more or less rounded grain of homogeneous micrite. The term itself implies no particular mode of origin. Most peloids are probably faecal pellets, strongly micritised (p.101) grains or are simply rounded fragments of reworked carbonate mud sediment. It is also possible that some peloids are produced directly by calcified algae and cyanobacteria or are cements formed with microbial involvement. Identifying the origin of these micritic particles in ancient limestones is rarely possible, hence the use of the non-genetic term 'peloid'.

An aggregate grain is formed when two or more originally separate particles become cemented together by micrite. In Recent environments the best known aggregate grains are the grapestones of the Bahama Banks. These are composed of rounded particles, initially bound together by microbial filaments which later become sites of cementation. At the same time the component particles are micritised. In ancient cemented limestones, it may be impossible in practice to distinguish between aggregate grains produced by a grapestone process and those which are reworked chunks of partly consolidated sediment.

Intraclasts are fragments of carbonate sediment that have been reworked within the basin of deposition. The classic example of an intraclastic carbonate as cited by Tucker & Wright (1990) is a mudflake conglomerate produced by reworking of desiccated sediment on tidal flats. Other intraclasts, produced by reworking of consolidated sediment within the depositional environment may also fulfil the definition of coated grain, peloid or aggregate grain.

Lithoclasts or extraclasts are fragments of cemented material from outside the immediate depositional area. These may be chunks of substantially older carbonates which have been subaerially eroded and transported to a shallow marine carbonate-producing environment.

As with the coated grains, some will not be identified with certainty. In these cases careful description is more important than seeking an instant label.

Carbonate Sediments and Rocks Under the Microscope

28 shows a thin section of loose grains from the pellet muds of the Great Bahama Bank. These grains, probably faecal in origin, show the typical ellipsoidal shape and, in addition to the clay-sized carbonate mud which makes up the bulk of them, they include a few slightly larger particles and are stained brown by organic matter. The speckled background is the epoxy in which the grains were embedded before sectioning. **29** shows a fine peloidal sediment which was associated with a lithified cyanobacterial mat (p.100). These well-defined structureless micrite grains are probably also faecal in origin. Note the uniformity in grain size.

30 shows a grainstone in which most of the grains are peloids, but with a variety of shapes and sizes. Many of the smaller grains are structureless, but some of the larger grains in the upper part of the photograph show some relict internal structure. These may be heavily micritised bioclasts. On the lower margin of the picture, to the left of centre, is a grain with a fine pattern of holes. This is characteristic of grains believed to be the faecal pellets of arthropods which have projections on the gut wall. They are often subrectangular in section. They are an exception to the definition of peloids as 'structureless'. By comparison with Recent material, they are interpreted as faecal in origin and they are therefore classified as peloids despite the internal structure. The smaller peloids in **30** are indeterminate in origin, although they seem rather too irregular in shape to be faecal pellets. Two micrite-walled foraminifera can also be seen.

Many peloid-rich rocks are characteristic of low-energy lagoonal environments and are likely to have a matrix of carbonate mud. This can mean that the peloids are difficult to distinguish, particularly if soft peloids have been compacted before lithification. **31** shows a finely peloidal rock in which individual peloids are, for the most part, distinct.

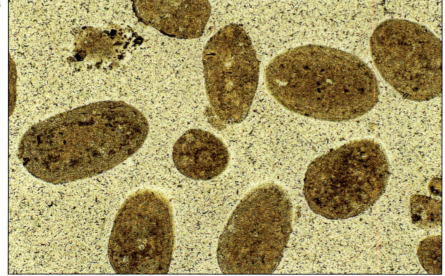

28 Unstained grain mount thin section, Recent, Bahamas, PPL, × 60.

Peloids, Aggregate Grains, Intraclasts and Lithoclasts

29 Unstained thin section impregnated with blue-dye-stained resin, Recent, Bahamas, PPL, × 90.

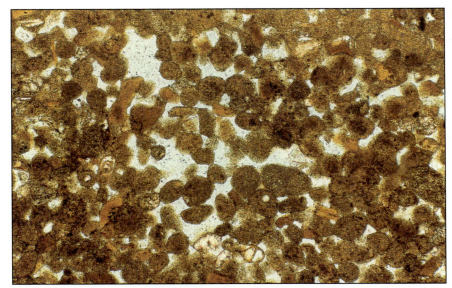

30 Unstained thin section, Upper Jurassic, Provence, France, PPL, × 32.

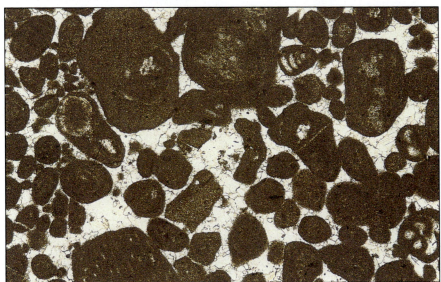

31 Unstained thin section, Upper Jurassic, Provence, France, PPL, × 38.

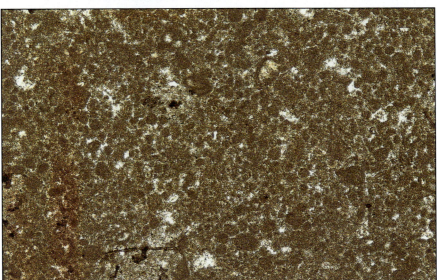

In **32** peloids are less distinct than in **31** and merge into carbonate mud. **33** shows very small (note the magnification), irregularly shaped micrite grains of uncertain origin. Some of these peloids have a faint structure and are perhaps heavily micritised bioclasts.

32 Stained thin section, Lower Carboniferous, Cumbria, England, PPL, × 38.

33 Unstained thin section, Lower Carboniferous, Lancashire, England, PPL, × 100.

Peloids, Aggregate Grains, Intraclasts and Lithoclasts

Peloidal cements are known from modern reef environments, and micro-organisms are thought to be at least partially responsible for their precipitation. On further diagenesis the peloidal texture can be destroyed; nevertheless, there are many records of finely peloidal fabrics in ancient reef sediments which are likely to be cements. 34 shows a fabric of poorly defined fine peloids of probable cement origin, often grading to slightly coarser crystals, from a cavity in a Carboniferous reef. Such fabrics are also considered under 'cements' (p.112).

35 shows an aggregate grain variety of an intraclast, comprising a number of separate particles, the centre upper of which is an echinoderm fragment, bound together by micrite. The surrounding sediment includes micritised bioclasts, peloids and coated grains.

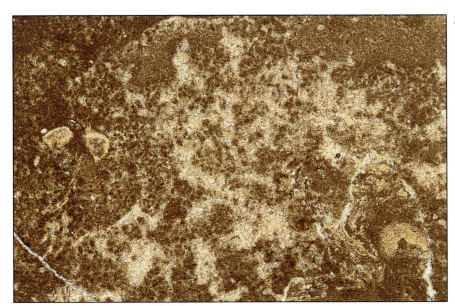

34 Stained thin section, Lower Carboniferous, Cumbria, England, PPL, ×34.

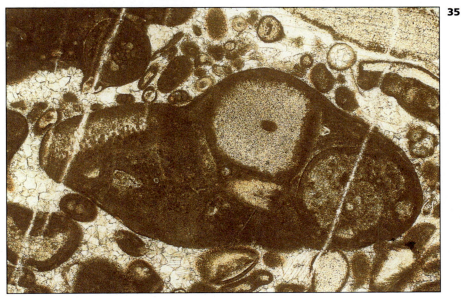

35 Unstained thin section, Lower Carboniferous, South Wales, PPL, ×35.

Carbonate Sediments and Rocks Under the Microscope

36 contains angular lumps of fine-grained sediment. These are pieces of partly lithified carbonate mud which have been eroded soon after deposition and locally reworked. They are thus classified as intraclasts. 37 contains a number of irregularly-shaped lumps of carbonate mud. Most are not sufficiently rounded to be classified as peloids and they are therefore best described as intraclasts, since they are likely to be locally reworked pieces of fine-grained sediment. This packstone also contains a number of foraminiferans and (at the top) an echinoderm fragment with a clear syntaxial rim cement (p.118).

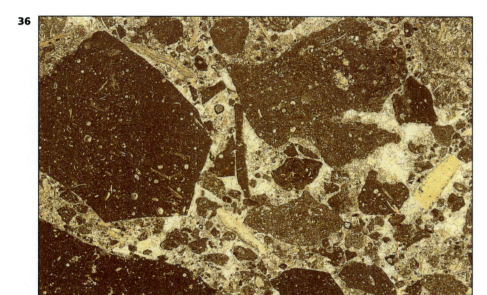

36 Unstained thin section, Lower Carboniferous, Derbyshire, England, PPL, × 14.

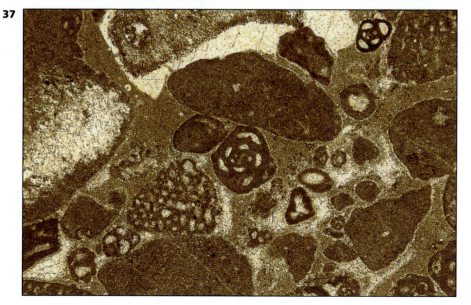

37 Unstained thin section, Lower Cretaceous, France, PPL, × 28.

Peloids, Aggregate Grains, Intraclasts and Lithoclasts

38 shows a grainstone with a variety of grain types including (across the centre of the photograph) a fenestrate bryozoan frond (p.66). However, the bryozoan has been coated in carbonate mud and so has been reworked from its original muddy, presumably low-energy environment, into this grainstone, although given its shape and susceptibility to breakage it is not likely to have come far. The grain thus classifies as an intraclast not a bioclast, although it could also be described as a coated grain. The photograph also shows ooids with a good radial structure and echinoderm fragments, the single crystals of which have developed twinning. The sparite cement shows two generations – an earlier very pale pink-stained non-ferroan calcite and a later blue-stained ferroan calcite.

38 Stained thin section, Lower Carboniferous, South Wales, PPL, × 25.

Carbonate Sediments and Rocks Under the Microscope

39–41 show examples of lithoclasts. In **39** an elongate fragment of oolitic grainstone has been incorporated in a younger sediment. This grain can be recognised as a lithoclast since both the ooids and their original cement have been truncated by erosion. In fact this is a piece of Carboniferous Limestone that was eroded and re-deposited in a Lower Jurassic carbonate sediment. In **40** a lithoclast is cemented in a bioclastic grainstone. The grain must have been lithified at the time of reworking to allow the rounding of the echinoderm fragment seen at the left-hand end of the clast. It is also of a more carbonate mud-rich lithology than the grainstone in which it is now incorporated.

39 Stained thin section, Lower Jurassic, South Wales, PPL, × 34.

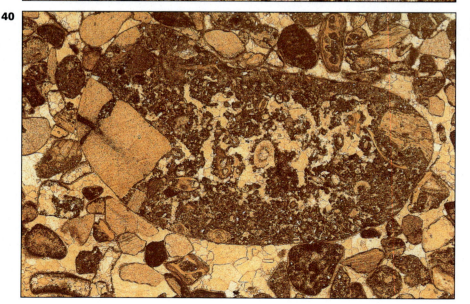

40 Stained thin section, Lower Carboniferous, Derbyshire, England, PPL, × 15.

Peloids, Aggregate Grains, Intraclasts and Lithoclasts

41 shows a limestone in which all the grains are compacted together so that there is no visible matrix or cement. These grains are lithoclasts of fine-grained Triassic and Jurassic carbonates eroded and re-deposited during the Tertiary. A well-rounded quartz grain can be seen lower left, and in the top left-hand corner the edge of a basaltic volcanic fragment occurs. Since the fragments are of subaerially eroded material this limestone could be also be regarded as a terrigenous clastic sediment, in which case it would be classified as a coarse sandstone or litharenite.

41 Unstained thin section, Tertiary, Mallorca, Spain, PPL, × 30.

BIOCLASTS

Of all the components of limestones it is the great diversity of bioclasts which the student of carbonate sedimentology is likely to find most daunting. Not only can there be great variety in one thin section, but the bioclast content of limestones varies with age, such that, for example, a bioclast grainstone of Palaeozoic age, consisting perhaps of crinoid, brachiopod and bryzoan fragments will look very different from one of Tertiary age, dominated by large foraminiferans, coralline algae and mollusc fragments, even though they may have been deposited in similar environments.

In trying to identify bioclasts the petrographer should keep in mind four things. Firstly, one should consider the size of the grain, and for this it is imperative to know the field of view of the microscope at each magnification. Secondly, the overall shape of the grain and whether it is complete or fragmented is important. Thirdly, and most importantly, it is vital to describe the wall structure. When looking at ancient carbonates the first step is to decide whether the wall structure is well preserved and therefore that it was of primary calcite mineralogy, or whether it is preserved as a mould or cast, and was therefore of primary aragonite mineralogy which has subsequently been altered. In the case of calcitic bioclasts it is then necessary to describe the wall structure by looking at the shape, size and orientation of the component crystals. This is often facilitated by viewing the section with polars crossed. Fourthly, the petrographer should always bear in mind the age of the rock being examined and thus the likely fauna and flora that will be encountered.

Young carbonate sediments, in which primary aragonite grains are more or less unaltered, can present particular problems because of the range of aragonite shell structures seen (particularly amongst the molluscs), which is absent in ancient limestones. In a book this size it is impossible to give comprehensive coverage of all bioclasts. We have tried to illustrate the most common bioclasts in limestones of various ages and for further information on invertebrates the reader is referred to the invaluable text of Majewske (1969): this book is full of useful drawings and photographs and is indispensable when dealing with aragonite structures in particular. Horowitz and Potter (1971), Flügel (1982) and Scholle (1978) are also well-illustrated texts which aid grain identification. There are a number of specialist books on fossil algae, including those by Johnson (1961), Wray (1977) and Flügel (1977).

Molluscs: Bivalves

Of the major molluscan groups, the bivalves are the most important contributors to the bioclast content of limestones, being present in marine environments since the Cambrian and in non-marine environments since the Carboniferous. However, it is in the Mesozoic and Cenozoic that they become major contributors to bioclastic sediments, being subordinate to the brachiopods during the Palaeozoic. Since the valves disarticulate easily on death of the organism, it is unusual to find complete shells, except perhaps of infaunal types which have not been disturbed. In most cases, it is single valves or fragments of valves which are found in limestones. Bivalves can grow to large sizes, so that in a standard thin section often only a small part of the shell is seen.

Bivalve shells can be wholly aragonite, wholly calcite or a mixture of the two, and shell structures also vary. Depending on what part of the shell a fragment comes from and the orientation of the section, individual fragments may not show the complete structure and mineralogy of the whole bivalve. The most important wall structures are foliated, prismatic, crossed-lamellar and homogeneous. Majewske (1969) includes an invaluable table summarising the shell mineralogy and structure of the major bivalve families.

42 is a photograph of a Jurassic limestone and shows four different bivalve shell structures. Dominating the left centre field of view is a transverse section of a shell with prismatic structure. Individual prisms with their polygonal cross-section can readily be seen. Along the top of the picture is a shell fragment with a foliated structure. In this example, the structure is quite irregular with bundles of calcite lamellae orientated in different directions. This is characteristic of oysters. Amongst oysters the foliated structure usually becomes more regular towards the inner part of the shell, suggesting that this is the upper surface of the valve in the photograph. These two shells with well-preserved structures are from dominantly calcite shells; there is no evidence in the photograph for the presence of other layers, although they may have been present in the original shell. At a low angle to the lower edge of the photograph is a shell fragment outlined by a thin dark line, probably a micrite envelope (p.101). Unlike the two previously described shells, this fragment does not have its original microstructure preserved. It is a cast of an originally aragonitic bivalve shell which has been replaced by sparry calcite, either by solution followed by later cementation, or by *in situ* recrystallisation (neomorphism, p.128). The fourth fragment visible in this photograph is the elongate fragment to the right of centre. Three shell layers are visible: along the lower side is a thin layer whose structure cannot be resolved at this magnification and above this is a foliated layer which shows a zig-zag pattern; finally, there is a replaced layer outlined by a thin dark line. This bivalve had mixed aragonite/calcite mineralogy originally. The foliated calcite layer is preserved and the original aragonite layer has been replaced by calcite.

42 Stained thin section, Middle Jurassic, England, PPL, × 40.

43 shows a fragment of a fairly thick-shelled bivalve, the structure of which is only faintly seen in this plane-polarised light view. However, in **44**, taken with polars crossed, the prismatic structure is visible, comprising crystals aligned roughly at right angles to the shell margins. At first sight **45** contains nothing that could be clearly identified as a shell fragment. However, to the right of centre there is a transverse section of a small group of calcite prisms and the rest of the sediment contains individual calcite prisms. These have been produced from the fragmentation of a prismatic bivalve shell. Break-up of prismatic shells with their coarser regular structure seems to occur much more readily than with other structures. Bivalves with a prismatic structure include genera such as *Pinna*, *Perna*, *Inoceramus* and some rudists. **45** also contains coiled pelagic foraminiferans and calpionellids (p.97).

43, **44** Stained thin section, Lower Cretaceous, Istrian Peninsula, × 34, **43** PPL, **44** XPL

Bioclasts

The bimineralic nature of some Quaternary bivalve shells is shown in **46**. This is a thin section stained with Feigl's Solution (p.8), which is used to distinguish aragonite from calcite. Aragonite takes on a rather 'spotty' grey or black colour and calcite remains unchanged. The two curved shells in this photograph are apparently two-layered, with a thick, inner layer of aragonite and a thinner, outer layer of calcite. Unfortunately, the stain masks the structure of the aragonite layer, although, in fact, the structure of both layers is homogeneous. In the left centre of the photograph is a fragment entirely of aragonite, and to the right is a fragment entirely made of calcite.

45 Stained thin section, Cretaceous, Isle of Skye, Scotland, PPL, × 45.

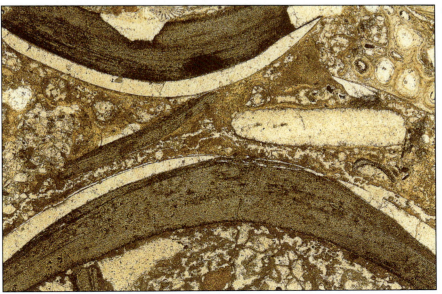

46 Thin section stained with Feigl's Solution, Quaternary, Morocco, PPL, × 14.

Carbonate Sediments and Rocks Under the Microscope

Crossed-lamellar structures are normally aragonite and comprise aggregates of branching lamellae which themselves are composed of small laths arranged at a constant angle to the lamellae. Although crossed-lamellar structures occur amongst the bivalves, they are more common in the gastropods. **47** and **48** are photographs of a Quaternary limestone containing unaltered aragonite mollusc fragments showing a crossed-lamellar structure. Some of these may be gastropods, but the shape of the straight fragment in the left part of the photographs suggests that it, at least, is from a bivalve. In an unstained section viewed with plane-polarised light, the structure will not always be evident, but in a stained section where different groups of lamellae have a different crystallographic orientation and take different amounts of stain (**47**), the distinctive structure is very clear. It is also distinctive with polars crossed, when spectacular patterns can be produced (**48**). The laths which make up the individual lamellae can just be seen in the upper part of the straight fragment in the left of the picture. This sediment also shows a radial fibrous aragonite cement of marine origin (p.104).

47, **48** Stained thin section, Quaternary, Barbados, × 21, **47** PPL, **48** XPL.

The term 'homogeneous' is applied to shell structures which comprise minute crystals of similar orientation, but which cannot be distinguished in viewing the grain with an ordinary light microscope. Such structures look clear with plane-polarised light, tend to stain an even, but fairly dense colour, and with polars crossed show an extinction shadow which moves through the fragment as the microscope stage is rotated. Apart from molluscs, the structure is found in trilobites, but there is unlikely to be confusion since trilobites are exclusively Palaeozoic and bivalves with a homogeneous structure are essentially Mesozoic and Cenozoic. 49 and 50 show a bivalve with a two-layered structure. The upper layer as seen in the photograph has been replaced and was therefore presumably originally aragonite. The lower layer has a homogeneous structure, the lines being striations rather than crystal boundaries. In this orientation the even interference colour in the view taken with polars crossed (50) suggests an almost single crystal structure, but, in fact, as the section is rotated an extinction shadow moves across the shell.

49, 50 Stained thin section, Middle Jurassic, England, × 28, **49** PPL, **50** XPL.

Carbonate Sediments and Rocks Under the Microscope

51 shows a section of a two-layered bivalve shell, an original foliated calcite outer part and a formerly aragonitic inner layer now replaced by ferroan calcite. The variation in thickness of layers is characteristic, with the inner layer thickening towards the umbo. The boundary between the two layers would meet the inner wall of the shell in the pallial line. The wavy fragment in the upper right of the photograph is a gently ribbed bivalve preserved as a cast.

52 is part of a thick-shelled oyster showing the foliated calcite structure. Most brachiopods also show a foliated calcite structure, but it is much more regular than that of the oyster shown here. The sediment is a fine peloidal and bioclastic limestone with detrital quartz grains.

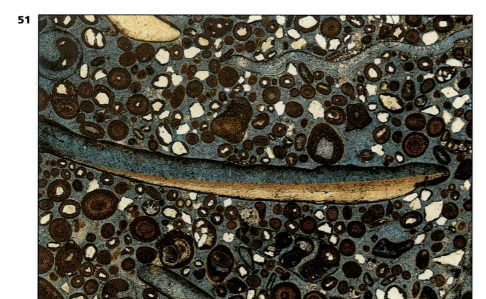

51 Stained thin section, Upper Jurassic, England, PPL, × 17.

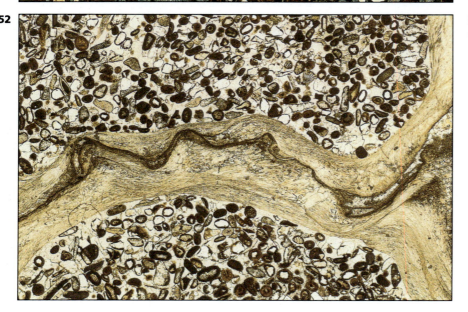

52 Unstained thin section, Jurassic, England, PPL, × 14.

Bioclasts

The rudists are a strange group of bivalves which become important in the Cretaceous. They include massive, bizarre forms where one valve was cemented to the substrate and a second acted as a 'lid'. Some had thick, but very porous walls. They were composed of calcite or a mixture of calcite and aragonite. 53 and 54 are sections of parts of rudist walls. 53 shows a laminated wall in which individual laminae have a prismatic structure, with prisms aligned at right angles to the laminae. In the upper part of the fragment, pores within the wall are filled with micritic sediment. 54 shows a rudist with a coarsely porous ('vesicular') wall. Parts of the wall are outlined by a thin, dark micrite coating and the pores are filled with sparite cement. The structure of the wall is not really evident in the photograph, but is, at least in part, prismatic.

53 Unstained thin section, Lower Cretaceous, Middle East, PPL, × 18.

54 Stained thin section, Lower Cretaceous, Istrian Peninsula, PPL, × 17.

Carbonate Sediments and Rocks Under the Microscope

In pre-Quaternary limestones, most originally aragonitic bivalves are preserved as moulds or casts. 55 and 56 show two curved bivalve shells that have suffered from solution and were therefore likely to have been originally composed of aragonite. In the view taken with polars crossed and the sensitive tint plate inserted, porosity shows up a violet colour. The upper shell, which shows thickening of the shell at the umbo (right), is a mould with a few small cement crystals lining it, especially at the left end, but the lower shell has been more completely replaced, although some porosity remains. 57 is a view of a section that has been impregnated with a dye such that porosity shows blue. The rounded fragment, probably of an abraded bivalve, has been completely dissolved, although there is a little fine spar cement on the lower side. The clear areas in the sediment are detrital quartz grains.

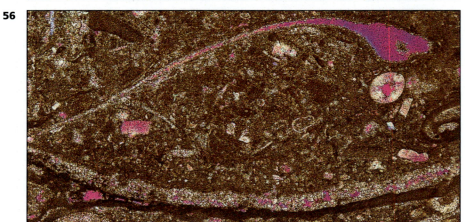

55, **56** Unstained thin section, Quaternary, Barbados, × 25, **55** PPL, **56** XPL with sensitive tint plate.

57 Unstained thin section impregnated with blue-dye-stained resin, Pliocene, Mallorca, Spain, PPL, × 20.

Bioclasts

In **58** the roughly rectangular areas outlined with dark micrite envelopes (p.101) are casts of molluscan fragments, probably bivalves. The original aragonite dissolved away and the resulting cavity (mould) was later filled with an equant calcite spar cement which is indistinguishable in morphology from the cement occurring between the grains. In **59**, however, although the shells are now made of coarsely crystalline calcite, some remnant of the original shell structure remains. This is best seen in the centre lower part of the photograph where lines of inclusions relating to the original laminated nature of the shell can be seen. This indicates that the original, presumably aragonite, shell did not completely dissolve leaving a void, but recrystallised *in situ* to calcite. The name given to this process is neomorphism (see also p.128). Other indications of neomorphism are the brown, inclusion-rich crystals in some parts of the shell fragments and the irregular-shaped crystals. Crystals in neomorphic fabrics also often show undulose extinction. It is also common for shell fragments to show clear neomorphic fabrics in some parts and ambiguous or cement-like fabrics elsewhere.

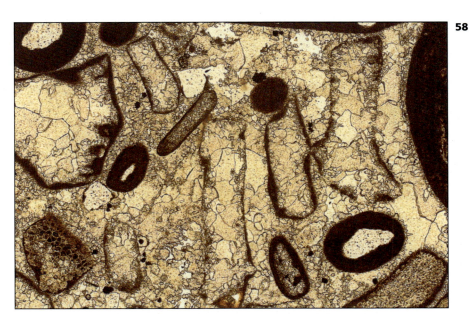

58 Unstained thin section, Mesozoic, Greece, PPL, × 40.

59 Stained thin section, Upper Triassic, England, PPL, × 40.

Carbonate Sediments and Rocks Under the Microscope

60 and **61** show thin-shelled pelagic bivalves from deep-water carbonate muds. These are important in Triassic and Jurassic basinal deposits and have a prismatic or homogeneous microstructure. In **60** the shells are straight or wavy, whereas in **61** curved shells have been stacked together. **62** illustrates a sediment composed almost entirely of pelagic bivalves, somewhat broken up by compaction.

60 Stained thin section, Middle Jurassic, Palencia, Spain, PPL, × 35.

61 Stained thin section, Triassic, Greece, PPL, × 40.

62 Unstained thin section, Triassic, Greece, PPL, × 40.

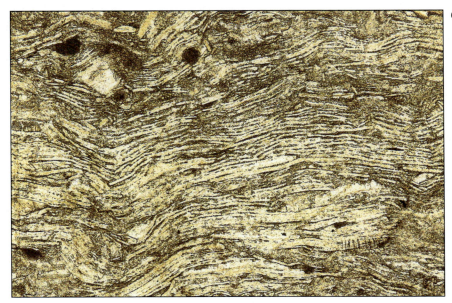

Molluscs: Gastropods

Gastropods are the second group of molluscs of major importance in limestones. Like the bivalves, they occur throughout the Phanerozoic, but are most abundant in Mesozoic and Cenozoic sediments. Gastropod shells are nearly all made wholly of aragonite, although there are a few with mixed mineralogy, comprising an outer layer of calcite and an inner layer of aragonite.

The most common gastropod shell structure is a crossed-lamellar structure like that of some bivalves. **63** shows a transverse section through a Pleistocene gastropod which is still aragonite and in which the crossed lamellar structure is visible. The body cavity is unfilled – a form of intragranular porosity (p.156).

63 Stained thin section, Quaternary, Morocco, PPL, × 30.

Because of the metastability of the aragonite shell, most gastropods are preserved as moulds and casts. **64** and **65** show respectively transverse and longitudinal sections through Carboniferous gastropod casts. In **64**, which shows a section almost identical to that in **63**, the inside of the shell is filled with fine sediment and the inside wall is clearly seen, whereas in **65**, the shell is only partly filled with sediment and the rest of the area is cement. Hence the inner wall is only seen near the aperture of the shell. **66** is a bioclastic wackestone in which the bioclasts are molluscan casts. The straight fragments are parts of bivalves and the chambered shell is a small gastropod. Foraminifera are sometimes confused with small gastropods, but the former are dominantly calcitic and rarely preserved as casts in limestones.

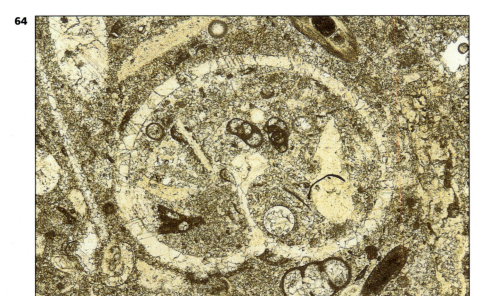

64 Unstained thin section, Lower Carboniferous, South Wales, PPL, × 35.

65 Stained thin section, Lower Carboniferous, North Wales, PPL, × 25.

Bioclasts

In **64–66** no trace of the original shell structure is visible; the casts formed by complete solution of the aragonite shell, followed by precipitation of cement into the void at a later date. However, as with bivalves, gastropod shells can also be neomorphosed to calcite and in such cases some trace of the original structure is retained. **67** is a photograph of a thick-shelled gastropod in which the original aragonite shell has been replaced by blocky calcite. In a number of places a trace of the original crossed-lamellar structure is still visible and towards the centre there is minor replacement by blue-stained ferroan calcite. The shell also has a geopetal infil (p.131) of fine bioclastic and terrigenous sediment.

66 Unstained thin section, Upper Jurassic, Provence, France, PPL, × 28.

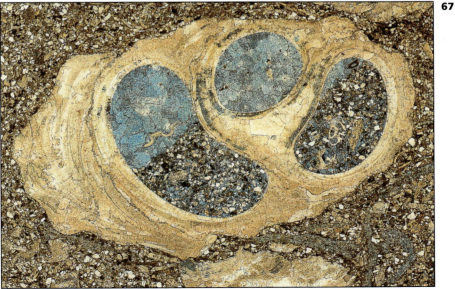

67 Stained thin section, Miocene, Mallorca, Spain, PPL, × 14.

Molluscs: Cephalopods

On the whole, cephalopods are not a major contributor to limestones. With the exception of belemnite guards and ammonite aptychi, they were entirely aragonite and are normally, therefore, preserved as casts in limestones. **68** shows a cast of a goniatite in cross-section. It is embedded in very cloudy, inclusion-rich radiaxial fibrous calcite cement (p.112), but with the exception of the body chamber, most of the chambers are filled with clear equant calcite. The positions of some septae are visible where they are coated with the earlier cloudy cement. A section of the coral *Hexaphyllia* (p.58) can be seen centre left.

69 contains sections of a number of small (juvenile?) ammonites, including sections parallel (left) and transverse (right) to the plane of coiling. Thin-shelled bivalves, probably pelagic, are also present.

68 Unstained thin section, Lower Carboniferous, Derbyshire, England, PPL, × 6.

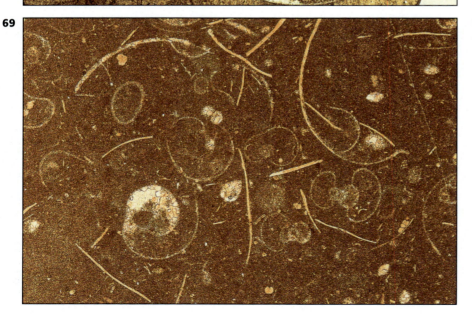

69 Stained thin section, Jurassic, Greece, PPL, × 40.

Bioclasts

70 and 71 are sections through a belemnite rostrum (guard). The view taken with polars crossed (71) shows an extinction cross which results from the radial orientation of individual calcite prisms that make up the rostrum. In this example the calcite has become twinned, the curved twin planes being clearly seen in the view taken with plane-polarised light (70). Twinning in sedimentary carbonates usually develops as a result of burial or tectonic stress (see also p.126).

70, **71** Stained thin section, Triassic, Greece, × 60, **70** PPL, **71** XPL.

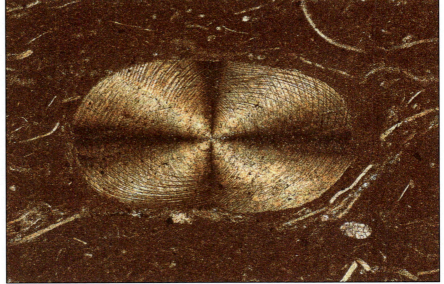

Brachiopods

Brachiopods are major contributors to the bioclastic content of shallow marine limestones, especially in the Palaeozoic. All articulate brachiopods (which are far more important than the inarticulates) have calcite shells and thus the primary wall structure is usually well preserved. The dominant component of brachiopod shells is a foliated layer consisting of fine fibres or prisms arranged with their long axes at a low angle to the length of the shell. Occasionally a thin outer layer of granular or prismatic calcite, with prisms arranged normal to the outside of the shell, is visible, but this is often not preserved in fossil specimens. There is a threefold division of brachiopods, based on shell structure: in punctate brachiopods the shell is perforated by small tubes; in others, known as pseudopunctate, the shell layering is interrupted by calcite rods or there are perturbations to the laminae; and in impunctate brachiopods both these features are absent and the shells are dense and imperforate.

Complete brachiopods, with both valves preserved, are occasionally encountered, although many are too large to be included in a single standard thin section. **72** shows a section through a small terebratulid brachiopod. The larger pedicle valve, smaller brachial valve and their attachment are clearly seen. The shell is embedded in carbonate mud, filled with a mixture of carbonate mud and blue-stained ferroan calcite cement and is cut by a vein of ferroan calcite. Terebratulids are punctate, but the punctae cannot be seen at this magnification.

The shell fragment in **73** is also a punctate brachiopod and shows the small tubular pores cutting the shell wall. **74** is a close-up of the wall of an impunctate brachiopod, with a fairly thick shell made of fine fibres aligned at a low angle to the shell wall. Although the shell is impunctate, its margins have suffered from micritisation (p.101), with some destruction of the shell by micro-borings. **75** shows a ribbed impunctate brachiopod encrusted on its upper surface by a bryozoan.

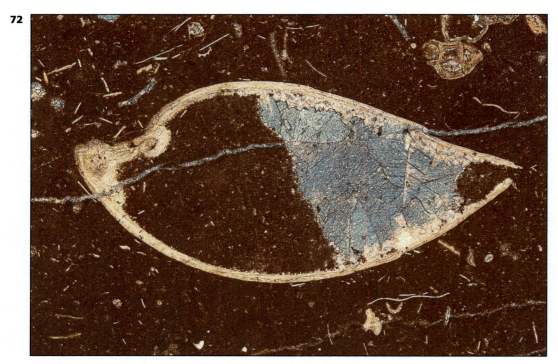

72 Stained thin section, Middle Jurassic, Palencia, Spain, PPL, × 30.

Bioclasts

73 Stained thin section, Middle Jurassic, England, PPL, × 35.

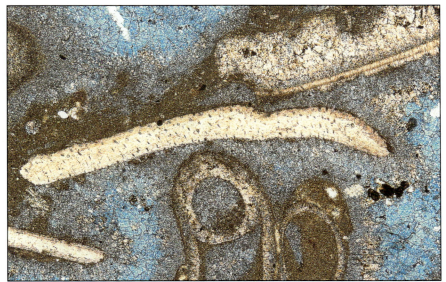

74 Stained thin section, Lower Carboniferous, South Wales, PPL, × 38.

75 Stained thin section, Devonian, Leon, Spain, PPL, × 40.

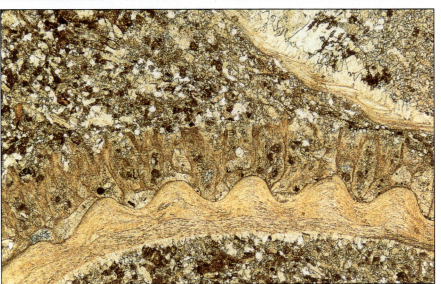

Carbonate Sediments and Rocks Under the Microscope

76 is a high-magnification view of a pseudopunctate shell. It shows well the laminated structure characteristic of the brachiopod shell and one of the calcite rods (taleolae) which fill the pseudopunctae. The inner surface of the valve is towards the top of the photograph. 77 shows a rounded fragment of a pseudopunctate brachiopod with the disturbances to the normal foliation, caused by the pseudopunctae, clearly seen. Individual laminae are deflected towards the interior of the shell, in this case towards the base of the photograph. If a pseudopunctate shell is sectioned such that the taleolae are cut almost at right angles to their length, the appearance is rather like a heavily knotted piece of wood (78).

76 Unstained thin section, Lower Carboniferous, Derbyshire, England, PPL, × 100.

77 Stained thin section, Lower Carboniferous, Derbyshire, England, PPL, × 35.

78 Unstained thin section, Lower Carboniferous, Derbyshire, England, PPL, × 38.

Bioclasts

Some pseudopunctate brachiopods have spines on one or both valves. These are hollow tubes, normally seen in cross-section (79, 80). Students sometimes confuse these with ooids to begin with, but the hollow centre and absence of any dark-looking micritic layers, plus the fact that they are tubes not spheres and usually show markedly elliptical as well as circular sections, serve to distinguish them. Like ooids, transverse sections of brachiopod spines show a pseudo-uniaxial cross with polars crossed. 81 shows an almost perfect longitudinal section through a brachiopod spine. 82 shows, unusually, a brachiopod valve (top) with its spines still attached and seen in various sections in the lower part of the photograph.

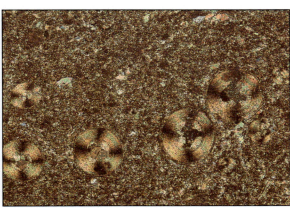

79, 80 Unstained thin section, Lower Carboniferous, Derbyshire, England, × 38, **79** PPL, **80** XPL.

81 Stained thin section, Lower Carboniferous, Derbyshire, England, PPL, × 20.

82 Stained thin section, Lower Carboniferous, Lancashire, England, PPL, × 18.

Carbonate Sediments and Rocks Under the Microscope

Although most sections of brachiopods show a dominantly foliated structure, there are some exceptions. If the fibres that make up the shell are sectioned transversely, a net-like structure with each crystal having a rectangular or rhombic shape is seen at moderate to high magnification. This is illustrated in **83**. A few brachiopods have a thick prismatic inner layer with the prisms aligned normal to the shell. **84–86** are pictures of a brachiopod that shows three layers. There is a thin outer layer (upper surface of shell as seen in the photograph) which is finely prismatic and can be best seen in the high-magnification crossed polars view (**86**). Beneath this there is a typical brachiopod foliated layer and a thick inner prismatic layer, the structure of which is most clearly seen in the low-magnification view taken with polars crossed (**85**).

83 Stained thin section, Lower Carboniferous, Lancashire, England, PPL, × 50.

Bioclasts

84–86 Unstained thin section, Lower Carboniferous, South Wales, **84** PPL, × 28, **85** XPL, × 28, **86** XPL, × 70.

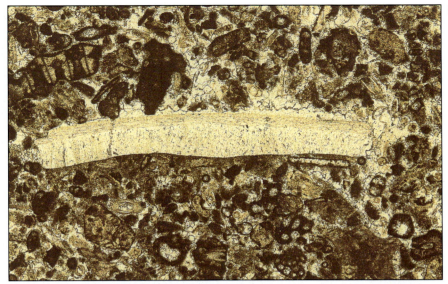

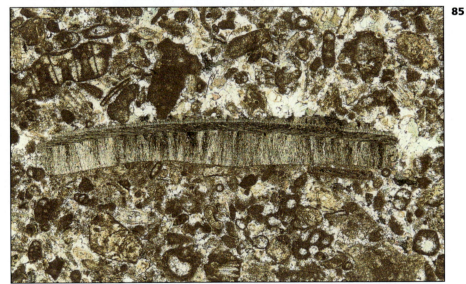

Corals

There are three major groups of corals: the Palaeozoic rugose and tabulate corals, and the Mesozoic and Cenozoic scleractinians. Scleractinian corals have hard parts composed exclusively of aragonite, but it appears that the rugose and tabulate corals were mostly or entirely calcite. Small fragments of coral can be difficult to identify but larger pieces can be recognised from their size and gross morphology.

The hard parts of all Recent scleractinian corals are made of radiating bundles of very fine fibrous aragonite. 87 and 88 are sections of *Siderastrea* from the Pleistocene of Barbados viewed at fairly high magnification to show the structure of the coral wall. The radiating bundles of fibres are more clearly seen in the view taken with polars crossed (88). The effect is to have many separate small areas

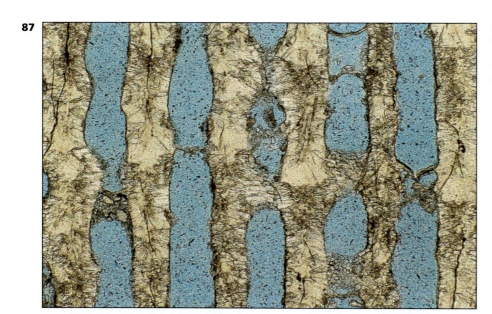

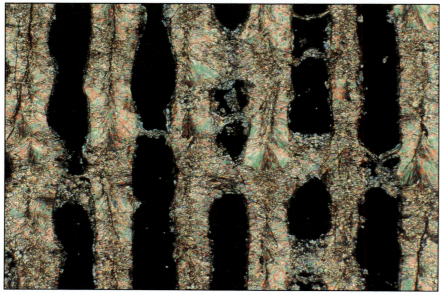

87, 88 Unstained thin section impregnated with blue-dye-stained resin, Quaternary, Barbados, × 90, **87** PPL, **88**, XPL.

of sweeping extinction as the stage of the microscope is rotated. **89** and **90** show a lightly cemented sediment consisting of subangular to rounded coral fragments. In small fragments the sweeping extinction pattern is not always visible.

89, **90** Stained thin section, Quaternary, Barbados, × 50, **89** PPL, **90** XPL.

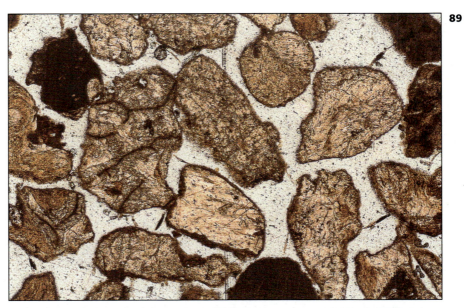

Carbonate Sediments and Rocks Under the Microscope

Older scleractinians are believed to have had the same structure, but being aragonite, they are usually preserved as moulds and casts and the original structure is lost. **91** is a photograph of a section of a colony of Jurassic corals. All the hard parts of the coral have been replaced by mauve-stained ferroan calcite and none of the original microstructure is visible. The morphology of the coral is only visible at the edge of the colony where internal sediment, which appears almost opaque in the photograph, filled the spaces between the septae prior to aragonite solution. In the interior of the colony, where little or no sediment infiltrated, the morphology is not apparent and there is no distinction between calcite occupying spaces formed from solution of the aragonite skeleton and that filling the primary pores between the septae. As with other aragonitic bioclasts such as gastropods and some bivalves, scleractinian corals can be preserved by neomorphism, in which some trace of the original wall structure remains during conversion to calcite. **92** is a photograph of a Jurassic scleractinian colony in which the wall structure seems well preserved. At higher magnification (**93**), traces of the original radiating aragonite fibres can be seen. However, the crystal boundaries (best seen right of centre) now delimit inclusion-rich calcite crystals.

91 Stained thin section, Jurassic, England, PPL, × 12.

92, 93 Stained thin section, Upper Jurassic, Morocco, PPL, **92** × 8, **93** × 25.

Tabulate and rugose corals, although constructed of calcite, appear to have a similar structure to the scleractinians, with little structure visible in plane-polarised light and irregular blotchy or sweeping extinction when the microscope stage is rotated with polars crossed. 94–96 are tabulate corals, 94 being a transverse section of a colony of the fasciculate tabulate *Syringopora*. The thick walls and absence of septae are characteristic. 95 and 96 are sections of the Silurian tabulate *Favosites*. In the low-magnification view (95), the corallite walls run vertically and the tabulae horizontally. The upper part of the colony is filled with sediment and the lower part with cement. 96 is an enlargement of the upper right part of the coral in 95 and it can be seen that the tabulae are very thin, appearing as little more than a dark line. The corallite walls have a similar dark line, but are fringed by a thin coating of clear calcite, in this case best seen in the upper part of the field of view where the chamber is filled with fine-grained sediment.

94 Stained thin section, Lower Carboniferous, Cumbria, England, PPL, × 20.

95, 96 Stained thin section, Upper Jurassic, Morocco, PPL, **95** × 7, **96** × 35.

97 shows sections of the corallites of a colonial rugose coral. Parts of transverse sections showing septae can be seen at the top of the picture and an oblique section showing the thin tabulae can be seen in the centre. Unstained patches on the coral wall are where there has been silicification.

A fourth group of corals occasionally encountered is the Palaeozoic heterocorals. 98 is a transverse section of the heterocoral *Hexaphyllia*. The shape and arrangement of septae is characteristic. The wall structure is finely granular to prismatic, most closely resembling that of an ostracod.

97 Stained thin section, Lower Carboniferous, Derbyshire, England, PPL, × 8.

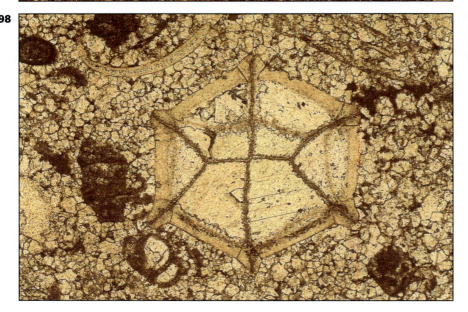

98 Unstained thin section, Lower Carboniferous, Cumbria, England, PPL, × 55.

Stromatoporoids

Stromatoporoids are colonial organisms, particularly common in Palaeozoic reefs and related sediments, although they re-appear in the Mesozoic. Most are large domed or tabular colonies, so that only a small part of an organism will be encountered in thin section. Most stromatoporoids have a fine rectangular wall structure which is visible in hand specimen, but the microstructure of the walls themselves is highly variable. It also appears that the walls were highly porous and are subject to alteration, such that it is difficult to distinguish primary and secondary structures. 99 and 100 are from a colony in which the rectangular structure of the skeleton was much more visible in hand specimen than in thin section. Nevertheless the feathery pattern of crystals revealed in the crossed polars view (100) is typical of rather altered stromatoporoid material.

99, 100 Unstained thin section, Silurian, Shropshire, England, × 28, **99** PPL, **100** XPL.

99

100

Sponges

Sponges include organisms whose walls are a mesh of spicules made of aragonite, calcite or silica or whose walls are massive calcite perforated by irregular canals. Some of the most common sponge remains seen in limestones are the calcite casts of formerly siliceous spicules. **101** shows an example of one of these with the characteristic spicule shape. **102** shows a number of circular areas which are transverse sections through sponge spicules formerly made of opaline silica. Some of these are now chert (low relief, unstained, slightly yellowish in colour), others have been replaced by carbonate, and some are a mixture of carbonate and silica.

101 Stained thin section, Lower Carboniferous, Cumbria, England, PPL, ×50.

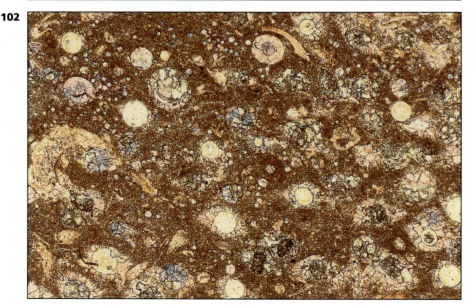

102 Stained thin section, Lower Carboniferous, Lancashire, England, PPL, ×50.

Bioclasts

103 is a stained section of a calcareous sponge with its red-brown stained calcite walls perforated by irregular canals. This sediment contains substantial detrital quartz and is cemented by blue-stained ferroan calcite. The Upper Palaeozoic colonial organism *Chaetetes*, once considered to be a tabulate coral, has been re-assigned to the sponges on the basis of similarity to living forms and the occasional occurrence of spicules. **104** shows a longitudinal section of *Chaetetes*, with thick, rather fibrous-looking walls. The morphological similarity to tabulate corals (p.57) and even solenoporoid algae (p.90) is clear. Small rhombic dolomite crystals can be seen in the upper part of the picture and yellowish dolomite fills much of the space within the organism in the lower part.

103 Stained thin section, Lower Cretaceous, Berkshire, England, PPL, × 20.

104 Stained thin section, Lower Carboniferous, Northumbria, England, PPL, × 18.

Carbonate Sediments and Rocks Under the Microscope

Bryozoans

Bryozoans are colonial marine organisms with a calcareous skeleton, and they are abundant on shallow shelves from the Ordovician to the present day. In the authors' experience, students are surprised by the abundance and diversity of bryozoans in carbonate sediments and this seems to reflect the rather cursory treatment they receive in some palaeontology courses, perhaps partly because they are too large to be microfossils, but in most cases, paradoxically, too small to study easily without a microscope, and also because they are not usually of value biostratigraphically. Bryozoans exhibit a wide variety of growth forms, including flat encrusting, upright, tubular branching and fan-like forms with a reticulate structure. Colonies consist of small tubes, called the zooecia, in which each animal lived. These are circular or polygonal in transverse section and may be partitioned. The wall structure of bryozoans is either laminated or finely granular. Larger fragments of bryozoans are usually quite easy to identify on size and gross morphology, but comminuted material may be difficult to distinguish from other material with a similar wall structure.

105 shows a longitudinal section through a bryozoan colony. The thin walls and larger zooecia in the central zone giving way to thickened walls and narrower zooecia in the outer zone is characteristic of some groups of bryozoans. The laminated nature of the wall can be made out in the thickened areas. Partitions within the zooecia are not evident in this example. At the top left of the photograph a transverse section of a smaller bryozoan can be seen and the bioclast to the lower left is a calcareous sponge. The zooecia are filled with a mixture of sediment and non-ferroan (pink-stained) and ferroan (blue-stained) calcite cements, and the colony is cut by some ferroan calcite-filled veins.

106 and **107** are sections of Silurian bryozoans. **106** shows thick laminated walls and internal partitions within the zooecia. The lower part of the colony in **107** shows that there are some circular structures within the zooecial walls. These are sections of acanthopores and are characteristic of some groups of bryozoans. **108** shows, on the left, a transverse section through a colony of upright bryozoans and, on the right, a bifoliate form, in which two lines of zooecia occur either side of a central wall.

105 Stained thin section, Lower Cretaceous, Berkshire, England, PPL, × 25.

Bioclasts

106 Stained thin section, Silurian, Shropshire, England, PPL, × 28.

107 Stained thin section, Silurian, Shropshire, England, PPL, × 14.

108 Stained thin section, Middle Jurassic, England, PPL, × 36.

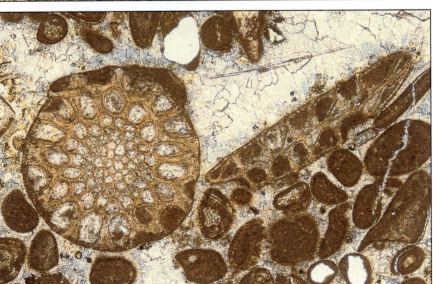

109 shows three lengths of bifoliate byozoan and, to the upper left, a bryozoan encrusting an originally bimineralic molluscan fragment. The sediment contains several molluscan fragments of different structures, as well as peloids, quartz grains and blue-stained ferroan calcite cement. **110** is an encrusting bryozoan, but the material encrusted has been completely replaced by calcite. In fact, the bryozoan was encrusting a scleractinian coral colony, but since the latter was aragonite, it has been dissolved, leaving a void to be later filled with mauve-stained ferroan calcite cement, whereas the bryozoan, made of calcite, did not dissolve. The colony is surrounded by carbonate mud.

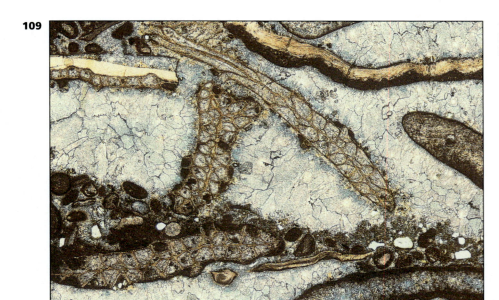

109 Stained thin section, Middle Jurassic, England, PPL, × 17.

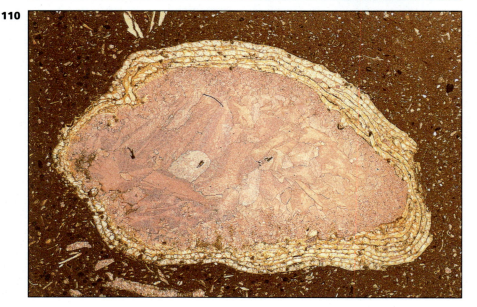

110 Stained thin section, Upper Jurassic, England, PPL, × 10.

Bioclasts

Another example of an encrusting bryozoan is *Fistulipora* which occupies most of the field of view in **111**, where part of it is attached to a thin shell fragment. It consists of convex-outward vesicular tissue between the zooecial tubes.

Majewske (1969) appropriately describes some bryozoans as appearing like 'lacy networks' and this is well seen in **112**. Fenestrate bryozoans often form erect fan-shaped colonies or fronds in which separate branches are linked by cross-pieces.

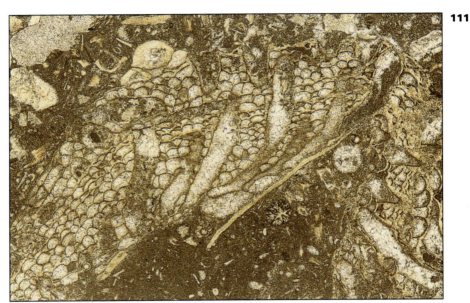

111 Stained thin section, Lower Carboniferous, Derbyshire, England, PPL, × 18.

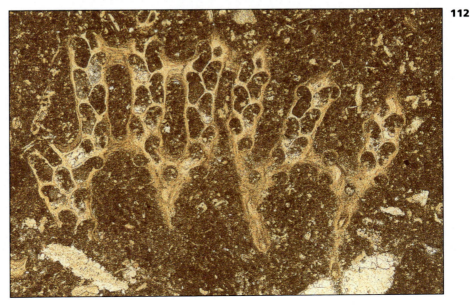

112 Stained thin section, Lower Carboniferous, Cumbria, England, PPL, × 32

113 shows a section of a fenestrate bryozoan frond running in a semicircle from the lower left to the lower right corner of the photograph. Each branch of the frond has a roughly circular cross-section. A small ladder-like section, also of fenestrate bryozoan, can be seen just to the right of centre of the picture. In detail, in well-preserved material, it can be seen that fenestrate bryozoans have a different wall structure from other bryozoan groups. In **114** two types of wall can be seen: an inner clearer part with an apparently spiny outer surface, and an outer darker laminated wall. In the large example in the left part of the field of view, structures ('spicules') can be seen to pass from the spiny projections of the inner wall through the outer wall. A further example of a fenestrate bryozoan frond, in this case in an intraclast, is well seen in **38**.

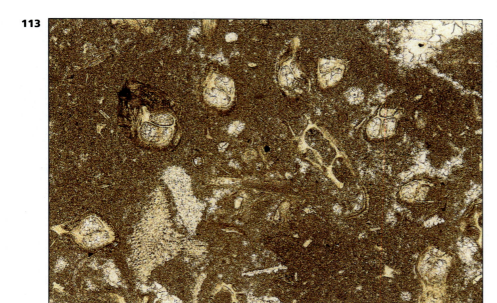

113 Stained thin section, Lower Carboniferous, Derbyshire, England, PPL, × 15.

114 Unstained thin section, Lower Carboniferous, Derbyshire England, PPL, × 48.

Foraminifera

The foraminifera are a group of single-celled benthonic or planktonic animals that range from the Cambrian to the present day. They are abundant, especially from the Late Palaeozoic onwards, and are important both for dating sediments and for palaeoenvironmental reconstruction. Foraminiferal tests show a huge variety of shapes and sizes, and the reader is referred to a textbook such as Brasier (1980) for a fuller description. Most foraminifera have several chambers and many are coiled. Differences include the number of chambers per whorl and the nature of the coiling. Difficulties in identification often arise because random sections through the same organism can look quite different.

The major foraminiferal groups also have quite different wall structures, which can therefore be an initial aid to classification. Most have calcareous tests (there are a few that are wholly organic and a very few with primary silica tests), and of these the vast majority are calcite so that wall structures of foraminifera are well-preserved in most limestones.

One wall structure, called agglutinated, characteristic of the Suborder Textulariina has individual grains from the sea-floor, such as other small bioclast fragments, micritic peloids and quartz grains bound together in a cement, which may be calcareous, ferruginous or organic. The foraminiferal test in **115**, *Pseudolituotubella*, is of this type; its dark micritic wall contains a number of small clear fragments of other bioclast debris. The aperture in this species (lower left of organism) is not simple, but shows a sieve-like structure.

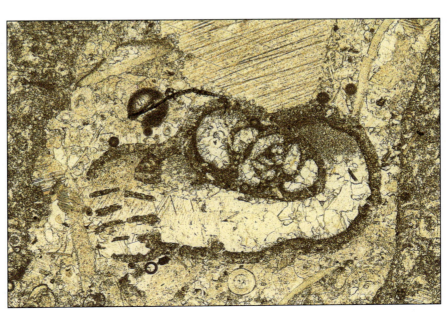

115 Unstained thin section, Lower Carboniferous, South Wales, PPL, × 35.

Carbonate Sediments and Rocks Under the Microscope

At first sight the foraminifera in **116** are difficult to pick out, because they have dominantly micritic walls and are embedded in micritic sediment. However, the chambers are filled with clear cement and the rather irregular 'knobbly walls' can be made out. As well as being agglutinated, the wall in these specimens is penetrated by a series of canals and is said to be 'labyrinthine'.

Amongst the larger foraminifera with agglutinated walls, the orbitolinids form a major component of the bioclasts in some Cretaceous sediments. **117** shows a typical section through an orbitolinid embedded in micritic sediment containing a substantial number of detrital quartz grains.

116 Stained thin section, Upper Jurassic, Western High Atlas, Morocco, PPL, × 35.

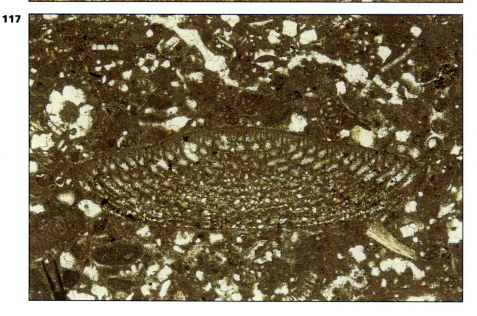

117 Unstained thin section, Lower Cretaceous, France, PPL, × 32.

Foraminifera belonging to the Suborder Miliolina have tests constructed of submicroscopic needles of magnesian calcite, usually in random orientation. The wall structure is described as porcelaneous, and in young sediments it has a very distinctive appearance, being yellow or brown when viewed with plane-polarised light and showing anomalous low birefringence with polars crossed. In older rocks this characteristic is lost and walls appear micritic. However, the Miliolina are dominantly a Mesozoic and Cenozoic group, and the other group with micritic walls, the Fusulinina, are almost entirely Late Palaeozoic. **118** is a peneroplid foram from a young sediment which shows the original brown colour very clearly. The fragments to the lower left of the foram are of coralline algae. This appearance is retained in Miliolina from a Pleistocene limestone (**119**). In **120**, from the Eocene, the Miliolina show some hint of the original structure and colour, and the walls have a less micritic appearance than those in many ancient sediments.

118 Unstained thin section, Quaternary, Barbados, PPL, × 50.

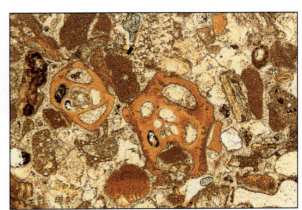

119 Stained thin section, Pleistocene, Mallorca, Spain, PPL, × 40.

120 Stained thin section, Eocene, France, PPL, × 36.

121 shows numerous sections of Miliolina having the typical preservation in an ancient limestone with thick micritic walls. This rock is lightly cemented and retains substantial porosity, clear in the photograph. One group of larger Miliolina is the alveolinids, in which the chambers are divided into many small chamberlets. **122** is a section of an alveolinid showing the numerous small round holes which are sections of the chamberlets. The sediment also contains many sections of Rotaliine foraminifera.

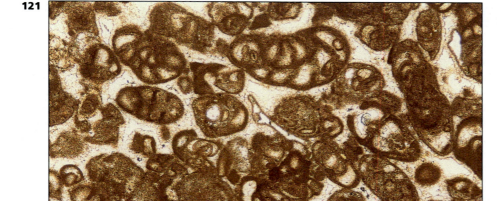

121 Stained thin section, Miocene, Mallorca, Spain, PPL, × 35.

122 Stained thin section, Eocene, Hungary, PPL, × 13.

Bioclasts

The Suborder Fusulinina includes foraminifera with a micritic wall, although many also have an inner fibrous layer. The Fusulinina are the dominant foraminiferal group of the Late Palaeozoic. The endothyraceans were small Fusulinina, often with a well-developed inner fibrous layer. **123** shows a number of foraminifera, including micrite-walled endothyraceans (just above centre and on the right-hand edge) and archaediscids. The latter are small Fusulinina with a fibrous outer wall and a thin micritic inner wall. In **123** two examples can be seen, one just below the centre of the photograph and one near the left-hand edge. In the photograph the fibrous nature of the wall is not clearly visible, but the contrast between the translucent outer wall and the dark micritic inner wall is evident.

Larger fusulines are characteristic of the Late Carboniferous and Permian and were mostly discoidal coiled forms. **124** shows several sections through these fusulines, with a transverse section at the top and various other tangential sections. The dark micritic wall is evident, but in two of them (one to the left and below, and one to the right and above the centre of the picture), the central chamber walls have been replaced by quartz, which appears translucent and very pale brown.

The problematic organism *Saccaminopsis* is regarded by some as a simple Fusulinina, but it has recently been re-assigned to the dasycladacean algae. It is illustrated in **156** (p.87).

123 Unstained thin section, Lower Carboniferous, South Wales, PPL, × 50.

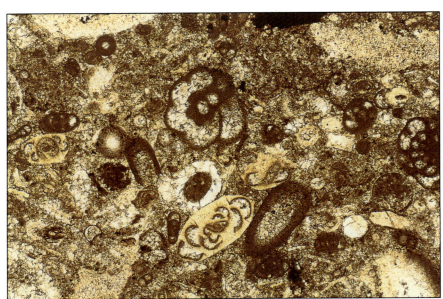

124 Unstained thin section, Permian, Middle East, PPL, × 13.

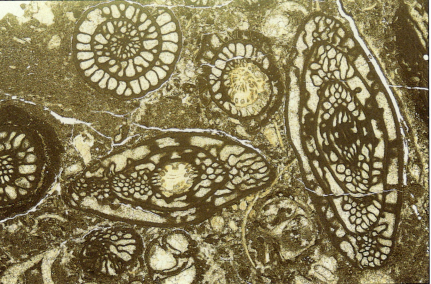

Carbonate Sediments and Rocks Under the Microscope

The largest of all the foraminiferal groups, the Suborder Rotaliina, comprises forms with glassy-looking (hyaline) walls which are more or less clear when viewed with plane-polarised light. The Rotaliines include some of the larger foraminifera which are especially common in the Early Tertiary and the planktonic foraminifera found in many Mesozoic and Cenozoic deep-sea carbonates.

One group of larger Rotaliines are the orbitoids (**125**, **126**), which range from the Late Cretaceous to Miocene. They are discoidal in shape with a characteristic arrangement of chambers, and include the readily identifiable discocyclinids (**125**). The hyaline wall structure is not clear in the example in **126**, but the chambers, including the embryonic central chamber, are well seen. The sediment contains abundant detrital quartz and blue-stained ferroan calcite cement fills all the original pore-space.

125 Unstained thin section, Palaeogene, Middle East, PPL, × 13.

126 Stained thin section, Palaeogene, Northern Spain, PPL, × 18.

Bioclasts

The rotaliaceans include the most famous of all the foraminifera, the nummulitids, which are particularly important rock-formers in Palaeogene rocks, where they may grow to sizes in excess of a centimetre in diameter. They are disc-shaped and have a hyaline wall composed of radial calcite, individual crystals being orientated with their long axes at right angles to the wall. **127** and **128** show nummulitids in sections which are both close to equatorial and to axial. The radial structure of the wall is especially well seen in the equatorial section (centre) in the view taken with polars crossed (**128**). Crystals are at extinction parallel to the edges of the photograph where the radial crystals are aligned N–S and E–W, giving a form of extinction cross. The foraminiferan in the upper left of the photograph is a discocyclinid.

127, 128 Stained thin section, Palaeogene, Mallorca, × 33, Spain, **127** PPL, **128** XPL.

Some Rotaliina have coarsely perforate walls and this is well illustrated in differently orientated sections in **129**. The foraminiferal chambers are filled with a mixture of smaller grains, carbonate mud and spar cement. A simple uniserial rotaliine is illustrated in **130**. The chambers are filled with a fine spar cement. The lower part of the photograph is dominated by a bryozoan.

129 Stained thin section, Eocene, Hungary, PPL, × 32.

130 Stained acetate peel, Miocene, Mallorca, Spain, PPL, × 48.

Bioclasts

The small planktonic globigerinacean foraminifera also belong in the Rotaliina and are illustrated in **131** and **132**. In **131** the larger examples with the ribbed ('keeled') tests are globotruncanids and the smaller rounded forms are globigerinids. **132** is a globigerinid limestone. There are a number of sections showing the arrangement of chambers and relatively thick, perforate test walls. The chambers are empty (a form of intragranular porosity, p.156), and much of the rest of the sediment consists of compacted thinner-walled examples which collapsed during burial of the limestone.

131 Stained thin section, Upper Cretaceous, Greece, PPL, × 70.

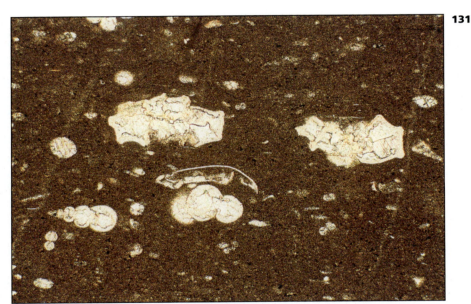

132 Stained thin section, Neogene, Cyprus, PPL, × 35.

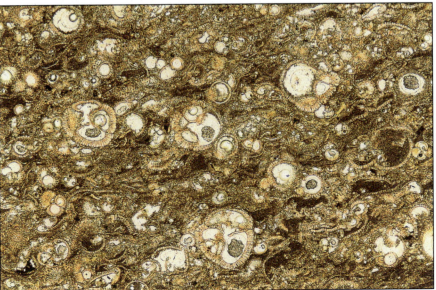

Echinoderms

Echinoderms are marine invertebrates that include forms attached to the sea-floor (mostly with stalks, e.g. crinoids, blastoids and cystoids), and crawling and burrowing forms (mostly echinoids). The calcite skeleton consists of plates of magnesian calcite permeated by a meshwork of organic material. The calcite in each plate, however, is secreted in the same crystallographic orientation so that it appears as a single crystal. It is this property which sets echinoderms apart from most other bioclasts and makes them amongst the easiest grains to identify in thin section. However, they often cause confusion with students, because of the wide variety of shapes and sizes of plates and the internal structure of the plate which may be visible if the spaces originally occupied by organic matter are filled with fine sediment. Echinoderm fragments are sometimes confused with calcite cement, especially since, in many echinoderm-rich rocks, cement precipitates syntaxially with the echinoderms, extending the original single crystal plate.

It is vital to be able to identify echinoderm fragments readily at an early stage of any course in carbonate petrography. They are present in almost every shallow marine limestone deposited in waters of normal salinity, and in many deeper-water carbonates too. Because most echinoderm material is disaggregated, it is often not possible to identify the original organism, although larger pieces of stalked organisms (e.g. blastoids and crinoids) and echinoid spines are often recognisable.

The most distinctive components of echinoderms are echinoid spines. These show a great variety in size and shape, but typically the calcite of each complete spine is in optical continuity so that they each appear as a single crystal. In limestones, evidence for the presence of the original organic material within the spine is present in the form of fine sediment or cement filling the spaces, and this normally has a radial appearance in cross-section. Something of the variety of echinoid spine morphology and ornament is seen in 133–140. 133–138 are transverse sections of spines, each showing a different type of radial ornament. 133 and 134 are from a Recent sediment and the porous nature of the spine, filled with blue-dye-stained resin, is well seen in the view taken with plane-polarised light (133). The homogeneous interference colour resulting from the single calcite crystal is seen in the photograph taken with polars crossed (134). 135 is from a small spine, probably with a hollow centre and pronounced external ribbing; it is embedded within some micritic sediment (dark) and cemented by ferroan calcite (stained blue). 136 shows a spine photographed with crossed polars and the surrounding black areas thus represent porosity in the rock. The spine shows pale interference colour in this orientation and the single crystal nature is again apparent from the homogeneous appearance. This spine is of similar stucture to that of 133 and 134, but the pores are filled with sediment and the spine appears to have undergone some abrasion.

137 shows the 'speckled' or 'dusty' appearance of most echinoderm plates when viewed with plane-polarised light. The specks are the areas occupied by organic material during life. The single crystal appearance is apparent in the view taken with crossed polars (138), where a fairly homogeneous interference colour is seen. The spine is cut by thin veins, but the vein filling is in optical continuity with the echinoid spine. Most echinoid spines have the calcite crystal c-axis parallel with their length, and transverse sections show interference colours lower than those often seen with carbonates, since, in sections of standard thickness, those cut nearly perpendicular to the c-axis of calcite show low-order interference colours.

Bioclasts

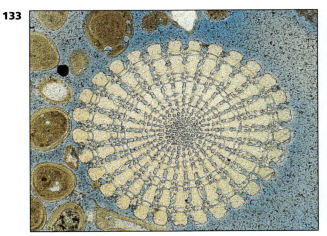

133, 134 Thin section impregnated with blue-dye-stained resin, Quaternary, Kuwait, ×50, **133** PPL, **134** XPL.

135 Stained thin section, Upper Jurassic, Dorset, England, PPL, ×70.

136 Stained thin section, Quaternary, Caribbean, XPL, ×38.

137, 138 Unstained thin section, Upper Jurassic, Provence, France, ×35, **137** PPL, **138** XPL.

77

Some sections of spines will be tangential and, therefore, elliptical in shape; occasionally longitudinal sections are seen and the radial nature of the ornament is not apparent. **139** and **140** are photographs of a longitudinal section through an echinoid spine, showing the ornament running along the spine. In this case the edge of the thin section passes through the lower right part of the photograph. The glass is, of course, isotropic and black in the view taken with polars crossed (**140**).

In the Palaeozoic, much of the echinoderm material in limestones is from crinoids, but, in section, plates show a great variety of shapes – round, rectangular, pentagonal and V-shaped, for example – depending on the part of the crinoid from which they are derived. Some coarser material is often recognisable as being from the echinoderm stalk, composed of a series of ossicles. **141** is a tangential section through a short length of crinoid stem comprising five ossicles, the ossicle to the left having been sectioned through its hollow centre; to the right, the section moves progressively towards the margin and the central hollow is missed.

142 and **143** show three echinoderm plates together. In plane-polarised light (**142**) the material looks homogeneous as though it might be a single plate, but with polars crossed (**143**) it can be seen that there are three plates, each a single crystal with homogeneous interference colours. The 'toothed' articulation between the plates is characteristic of some crinoid material.

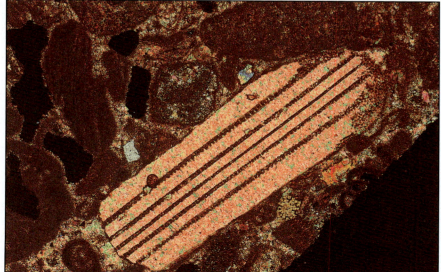

139, **140** Stained thin section, Pleistocene, Mallorca, Spain, × 55, **139** PPL, **140** XPL.

Bioclasts

141 Stained thin section, Lower Carboniferous, Derbyshire, England, PPL, × 15.

142, 143 Stained thin section, Lower Carboniferous, Derbyshire, England, × 35, **142** PPL, **143** XPL.

Carbonate Sediments and Rocks Under the Microscope

In **144** the crinoid plates are impregnated with opaque material (iron oxide) and this shows the sieve-like structure of the original plate. The iron oxide has filled the pores previously occupied by organic material, although in this case it is likely that the oxide has begun to replace the surrounding calcite as well. **145** shows a section of a typical Upper Palaeozoic crinoidal limestone, more than half of which consists of speckled crinoid plates surrounded by some clear cement which has grown syntaxially (p.118) on the single crystal echinoderms. The other common bioclasts in this section are fragmented fenestrate bryozoans.

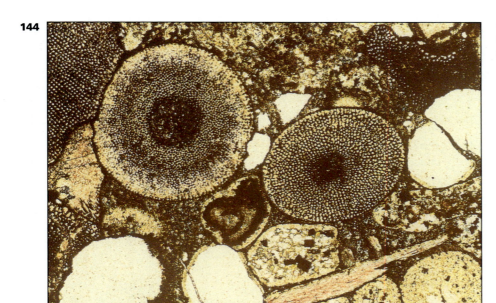

144 Stained thin section, Carboniferous, Central High Atlas, Morocco, PPL, × 55.

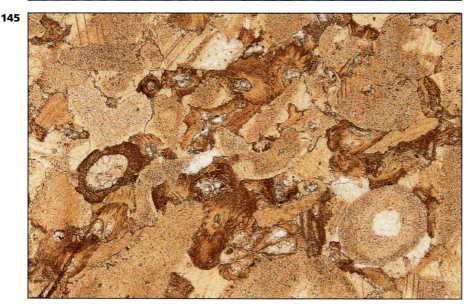

145 Stained thin section, Lower Carboniferous, Derbyshire, England, PPL, × 35.

Calcareous algae

The calcareous algae are, perhaps, the most difficult of all bioclasts, embracing a huge variety of shapes and structures and including many organisms of uncertain affinity. Sometimes it seems that algae are used as a dustbin for all bioclast material that cannot otherwise be identified. Like the bryozoans, calcareous algae get a raw deal in most university palaeontology courses and the description of them has often been left to carbonate sedimentologists, particularly since it is usually based on a study of thin sections. In this book we can briefly introduce some of the common types of calcareous algae. Other sources of information on fossil algae include Johnson (1961), Wray (1977), Flügel (1977) and Brasier (1980). Flügel (1982) is a useful source of further references on algae of different ages.

Despite the problems in their recognition and identification, calcareous algae are abundant in many shallow marine limestones and can be very sensitive to changes in water depth and energy, light penetration, etc. They are therefore useful in reconstructing palaeoenvironments.

The benthonic calcareous algae belong to two major groups: the Chlorophyta (green algae) and Rhodophyta (red algae). The minute planktonic Chrysophyta, the coccolithophorids, are too small to distinguish in thin section. They are studied by using an electron microscope and are thus beyond the scope of this book. Some organisms previously assigned to the blue-green algae (Cyanophyta) are now regarded as cyanobacteria and are included here under calcified microbial filaments (p.99). Soft-part morphology is critical in recognizing major groups of living algae and since these are not preserved in fossil material, there can be major problems in classifying algae from the geological record.

As well as contributing to the bioclastic content of shallow marine limestones, the calcareous algae may have a vital role in the production of carbonate mud sediment. In some modern environments carbonate mud is derived from the disintegration into component micron-sized crystals, of aragonite encrustations around the algae.

There are two major groups of calcareous green algae: the dasycladaceans and the codiaceans. The dasycladaceans are erect, rooted plants with a central stem and many branches, usually arranged at right angles to the stem. The calcification, which is in the form of aragonite needles a few microns long in modern examples, forms a sheath around the stem and through which the branches pass. Fossil material, which is usually either micritic calcite or a sparry calcite cast of the original, consists of a hollow cylinder perforated by tubes which represent the position of the branches.

Carbonate Sediments and Rocks Under the Microscope

One characteristic Carboniferous dasycladacean is *Koninckopora*, seen in **146** and **147**. *Koninckopora* is usually well preserved and it seems likely that, unusually for a dasycladacean, it was of primary calcite mineralogy. **146** shows, on the left, a perfect transverse section through *Koninckopora* showing the central cavity, formerly occupied by the stem and, in this case, the rather thin zone of calcification with moulds of the branches. To the right of this is a not quite complete section, at a slight angle to transverse; it shows the honeycombed appearance that is characteristic of fragments of *Koninckopora*. **147** shows some short longitudinal sections of *Koninckopora* in a finely bioclastic and peloidal limestone.

146 Unstained thin section, Lower Carboniferous, South Wales, PPL, × 28.

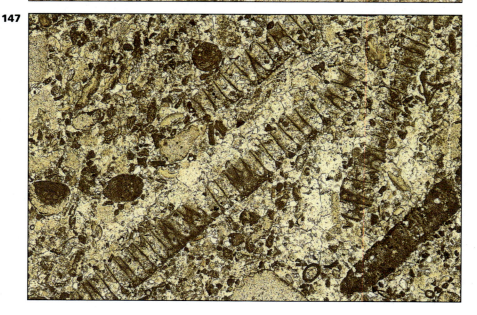

147 Unstained thin section, Lower Carboniferous, Cumbria, England, PPL, × 24.

Bioclasts

In **146** and **147,** the calcified areas of the algae are preserved as micrite. If original aragonite is dissolved or replaced by coarser sparry calcite and if the pores are not filled by sediment, much of the structure of the plant will not be visible. **148** shows several transverse sections of probable dasyclads and a more longitudinal section (lower left). The hollow centres are visible, but there is no clear evidence for the branches. In **149**, there are short broken lengths, which are probably sections of the stem of a dasycladacean, associated here with micritic peloids and cemented by sparry calcite.

148 Unstained thin section, Upper Jurassic, Provence, France, PPL, × 50.

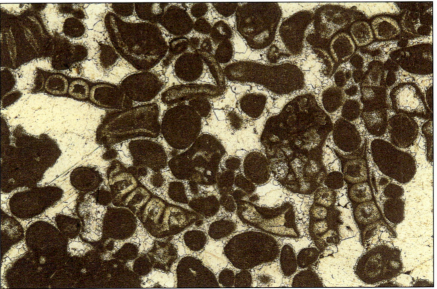

149 Unstained thin section, Upper Jurassic, Provence, France, PPL, × 28.

Carbonate Sediments and Rocks Under the Microscope

Codiacean algae are mostly erect segmented forms. Each segment consists of filaments embedded in the calcifying carbonate, which in Recent forms is fine-grained aragonite. Codiaceans usually break down into the individual segments and lack the central cavity, previously occupied by the stem, found in the dasyclads. The best-known Recent codiacean is *Halimeda*, abundant in the shallow carbonate environments of the Caribbean. **150** shows a section of a *Halimeda* segment now incorporated in a beachrock. The brown colour of the calcifying material is typical of Recent examples, as are the small tubes normal to the surface at the margins of the segment, with larger tubes in the central area. **151** shows a number of *Halimeda* segments in a lightly cemented sediment. The calcifying aragonite is very dark in colour, and in giving enough light to show the structure of the segments, the rest of the photograph is overexposed. The dark colour is a result of the organic matter within the plate.

150 Stained thin section, Quaternary, Caribbean, PPL, × 36.

151 Unstained thin section, Quaternary, Barbados, PPL, × 40.

Bioclasts

An important group of extinct organisms attributed to the codiaceans is the phylloid algae. Phylloid means 'leaf-like' and these algae occur as thin curved plates in build-ups and associated sediments of Late Palaeozoic age. They can only be easily identified where sediment has filled pores at the margin of the plate. The calcifying material, being aragonite, is usually replaced by sparry calcite, so that poorly preserved material can closely resemble other formerly aragonite bioclasts such as molluscs. 152 and 153 show sections of phylloid algae. In 152 a number of plates are seen, with sparry calcite centres and fine sediment filling the marginal pores. This feature is more clearly seen in another example viewed at higher magnification (153). In this case the phylloid algal plate is preserved in a porous spar-cemented sediment.

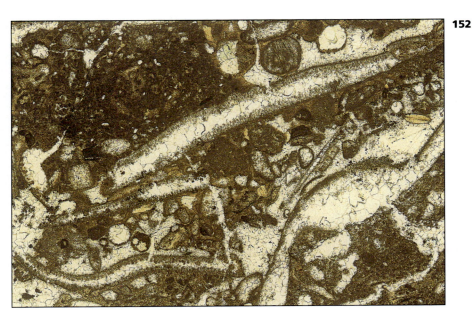

152 Unstained thin section, Upper Carboniferous, Spitsbergen, PPL, × 14.

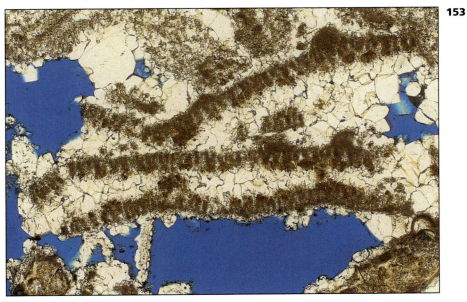

153 Thin section impregnated with blue-dye-stained resin, Upper Carboniferous, Spitsbergen, PPL, × 38.

A number of problematic groups are often attributed to the green algae. These include calcispheres – small, hollow, usually micritic-walled spheres found in shallow-water, low-energy limestones, especially in the Late Palaeozoic. **154** shows a number of bioclasts with hollow circular sections, but displaying a variety of wall structures: these are calcispheres.

A second group attributed by some to the dasycladaceans, although superficially they bear little resemblance, are the beresellids and palaeoberesellids. These are important rock-formers in some Upper Palaeozoic limestones and consist of sparsely branching hollow cylinders with some internal partitions, a fraction of a millimetre in diameter. Some workers have attributed these organisms to sponges or foraminifera. **155** shows a limestone composed almost entirely of palaeoberesellids. A short longitudinal section can be seen in the centre of the field of view and the rest of the rock comprises transverse and tangential sections cemented by sparry calcite. Palaeoberesellids may be confused with finely comminuted echinoderm debris, since they both show a speckled appearance and comprise single crystals, although palaeoberesellids often show undulose rather than sharp extinction. Like echinoderms, palaeoberesellids in grainstones commonly develop syntaxial cements and this has happened in **155**. Their small size and characteristic shape is sufficient to distinguish palaeoberesellids in most cases.

154 Stained thin section, Lower Carboniferous, Derbyshire, England, PPL, × 52.

155 Unstained thin section, Lower Carboniferous, South Wales, PPL, × 100.

Bioclasts

Another possible dasycladacean is the organism *Saccaminopsis*, once regarded as a foraminiferan. It usually consists of circular or elliptical casts, often with a 'tail' at one side. Typical sections of *Saccaminopsis* can be seen in **156**. Individual casts may formerly have been linked to give a segmented plant, and the 'tail' seen on the example on the right-hand side of **156** and the broken end of the example on the left may be evidence of the 'join' between segments.

Another group of algae generally referred to the green algae are the charophytes. These can be quite tall upright plants with numerous branches, but are confined to fresh or brackish water environments, although their calcified parts may be carried to other environments. Sometimes the stem is calcified, and the characteristic cross-section is seen in **157**, with its central cell cavity surrounded by smaller cortical cell cavities. However, it is the female reproductive parts of charophytes that are more normally calcified. These consist of a hollow sphere or ellipsoid of calcite, often with a spiral ornament on the outside, and are called oogonia or gyrogonites. Cross-sections of charophyte oogonia embedded in carbonate mud can be seen in **158**, along with thin, curved ostracod valves.

156 Stained thin section, Lower Carboniferous, North Wales, PPL, × 18.

157 Unstained thin section, Lower Cretaceous, Vercors Massif, France, PPL, × 50.

158 Unstained thin section, Upper Jurassic, Provence, France, PPL, × 25.

The other important group of calcareous algae are the red algae, comprising three families of particular geological significance: Corallinaceae, Solenoporaceae and Gymnocodiaceae. Of these, the coralline algae are the most important, being especially abundant in Cenozoic reefs and associated environments. The coralline algae exhibit many different external forms, such as encrusting, nodular and branching, segmented types. All have an internal arrangement of two different sizes of cells, although both may not always be visible in fragmented material. The cells are small and can be missed if the grain is only examined with low magnification. **159** is a low-magnification view of several coralline algae. The photograph shows transverse sections and one longitudinal section through the plant. **160** shows a higher magnification view of a rounded coralline algal fragment, showing the two different cell arrangements. The micritic and fibrous cement crusts in this sample are further illustrated and described in **194** (p.105).

159 Unstained thin section, Quaternary, Barbados, PPL, × 17.

160 Stained thin section, Quaternary, Rodriguez Island, Indian Ocean, PPL, × 28.

Bioclasts

The term rhodoid or rhodolith is used for roughly spherical, unattached nodules dominantly composed of coralline algae (some geologists classify them as coated grains – see p.22). **161** is a section through a complete rhodoid, the dark red-brown stained areas being the algae. Note that other organisms are included within the structure. Some of the holes within the alga are sporangia, others may be borings. **162** is an enlargement of the upper part of the rhodoid in **161**, showing the fine cellular structure of coralline algae. A quartz grain is lodged within the algal colony (upper right) and the top surface has an attached discocyclinid foraminifer (p.72).

161, 162 Stained thin section, Miocene, Mallorca, Spain, PPL, **161** × 7, **162** × 18.

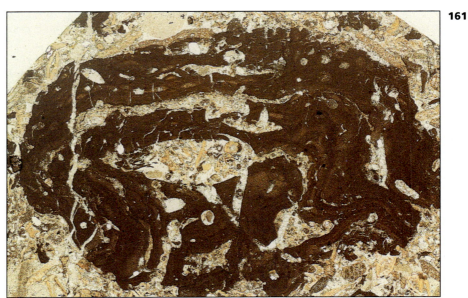

161

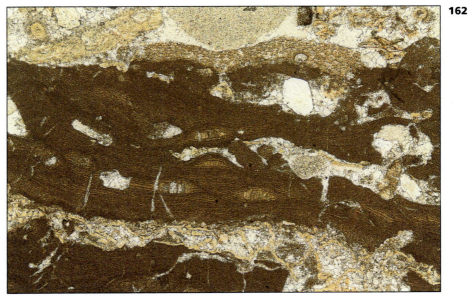

162

The solenoporoid algae are longer ranging (Cambrian–Miocene) than the corallines, but less widespread. They are composed of encrusting, nodular or occasionally branching masses. **163** shows the internal structure of a solenoporoid alga comprising a mass of tubes which are circular or polygonal in cross-section. The colony is seen in transverse section in the bottom left of the picture, whereas the upper part shows almost longitudinal sections in which occasional cross-pieces can be seen. There is a potential confusion between solenoperoids, tabulate corals and probable sponges like *Chaetetes*, especially if material is fragmented. **164** is a section through a branching colony. The characteristic cellular structure is preserved around the margins, but the centre part of the colony has been recrystallised.

The third group of red algae, the gymnocodiaceans, are of only local importance, most descriptions being from Permian or Cretaceous rocks. They are segmented plants, resembling the dasycladaceans in morphology in that they comprise a central, sometimes weakly calcified, stem surrounded by a sheath of carbonate through which branches pass. Compared with the dasyclads, the gymnocodiaceans have much smaller diameter branches which are oblique, rather than normal, to the margin of the segment. **165** is a section of a Cretaceous bioclastic wackestone in which most of the bioclasts are gymnocodiaceans, seen here in both transverse and longitudinal sections. Gymnocodiaceans secreted aragonite and hence the original calcareous parts are replaced by sparry calcite. The structure of the plant is visible because fine sediment has filled the spaces originally occupied by the soft parts. In the longitudinal section (centre left) the small oblique branches are well seen. In places the fine carbonate filling the centre of the stem has retained a slight impression of an original filamentous nature.

As with the green algae, there are some groups included with the red algae by some scientists, which are regarded by others as not being algae at all. There is a group of cylindrical branched, segmented microfossils, usually with a poorly preserved internal structure, but apparently having walls of microgranular or finely fibrous carbonate, which include the genera *Komia*, *Stacheia* and *Ungdarella*. Both longitudinal and transverse sections of these organisms can be seen in **166**, in a rock that is composed almost entirely of these organisms. They are particularly common in Upper Palaeozoic rocks and are regarded by some workers as stromatoporoids and by others as ancestral red algae. In plane-polarised light, poorly preserved material may be confused with echinoderms, but with polars crossed, as in **166**, it can be seen that they are not composed of single crystals. Some similarity with the gymnocodiaceans of **165** is apparent in this view.

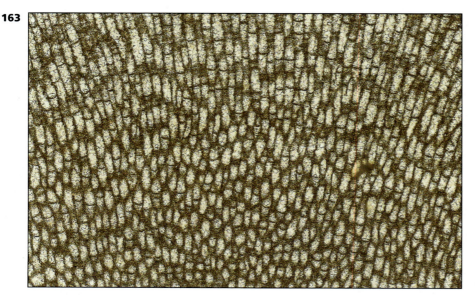

163 Unstained thin section, age and locality unknown, PPL, × 42.

Bioclasts

164 Unstained thin section, Lower Carboniferous, Poland, PPL, × 13.

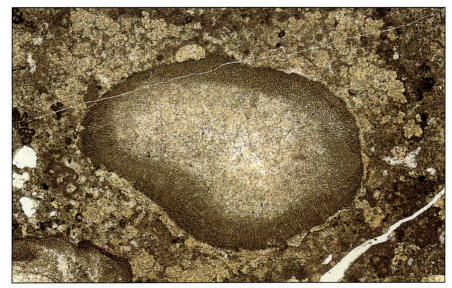

165 Stained thin section, Lower Cretaceous, Tunisia, PPL, × 35.

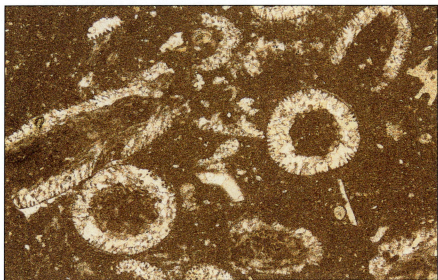

166 Unstained thin section, Lower Carboniferous, South Wales, XPL, × 30.

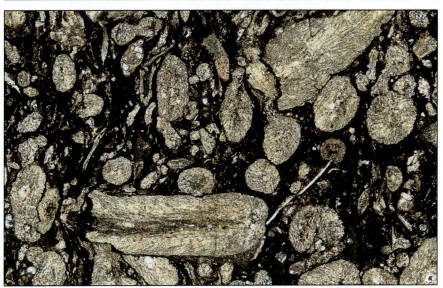

91

Arthropods: Trilobites

Only two major groups of arthropods have calcified skeletons: the trilobites and the ostracods. Trilobites are an exclusively marine Palaeozoic group, especially common in Cambrian rocks, but locally common in younger Palaeozoic strata. The calcified skeleton is composed of fine-grained calcite. Fragments are clear (sometimes slightly yellowish) in plane-polarised light, and normally take an even, bright stain in stained thin sections. Under crossed polars, areas of extinction move across fragments as the stage of the microscope is rotated. The structure appears identical to the homogeneous structure of some bivalves (p.37). 167 shows a cross-section of a trilobite skeleton. The arched appearance and the incurved margins giving a hook (or 'shepherd's crook') shape are characteristic and are often seen in fragmented material. The absence of any structure when viewed in ordinary light is also characteristic. 168 is from a limestone in which most of the larger fragments are from trilobites. Two examples of the incurved margin are also well seen.

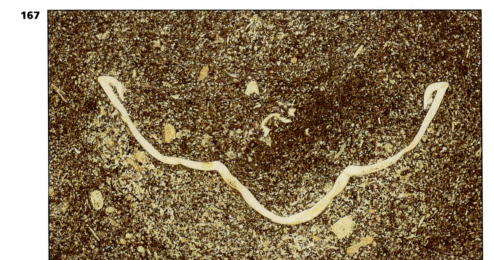

167 Unstained thin section, Lower Carboniferous, South Wales, PPL, × 8.

168 Unstained thin section, Lower Carboniferous, Derbyshire, England, PPL, × 38.

Bioclasts

169 and **170** show part of a large trilobite fragment. In the view taken with plane-polarised light (**169**) the apparently structureless nature can be seen. With crossed polars (**170**), areas of extinction can be seen within the skeleton. As the stage of the microscope is rotated these extinction zones would move across the skeleton.

169, 170 Stained thin section, Devonian, Leon, Spain, × 28, **169** PPL, **170** XPL.

Arthropods: Ostracods

Ostracods are widespread and often abundant bivalved arthropods, mostly a couple of millimetres or less in size. The two valves are of the same shape, but of slightly different size and can sometimes be seen to overlap. Since the valves are shed as the organism grows, layers of disarticulated valves in limestones are more common than complete shells. The shells are calcite and retain a homogeneous or finely prismatic microstructure. **171** shows a group of complete ostracods in cross-section. The two valves can be seen, and in a number of cases, the characteristic overlap of one valve by another is also seen. **172** shows a carbonate mudstone in which the only visible bioclasts are disarticulated and sometimes fragmented ostracods, some of which have been fractured during compaction.

171 Stained thin section, Lower Carboniferous, Cumbria, England, PPL, × 40.

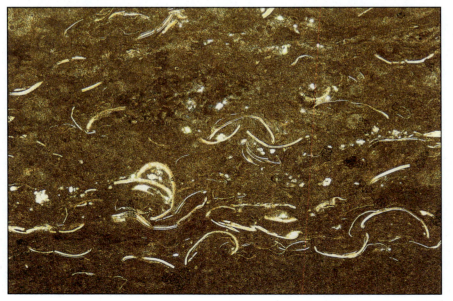

172 Stained thin section, Miocene, Mallorca, Spain, PPL, × 35.

Bioclasts

173 and **174** show a high magnification view of a single articulated ostracod filled with coarse calcite cement. The slightly different size of the two valves is evident. The finely granular or prismatic structure of the ostracod valve results in extinction in the north, south, east and west positions, and this can be seen in the view taken with polars crossed (**174**).

173, **174** Unstained thin section, Lower Carboniferous, Lancashire, England, ×90, **173** PPL, **174** XPL.

Worm tubes and vermiforms

Organisms referred to as calcareous worm tubes are known from throughout the Phanerozoic. Many of the records of worm tubes, particularly of Carboniferous serpulids and spirorbids, have been reinterpreted as vermiform gastropods. It has further been suggested that these are not gastropods and may in fact be more closely related to *Tentaculites* (see below). To the non-specialist, this leaves the situation in a degree of confusion. However, calcareous worm tubes undoubtedly occur in carbonate sediments and may be of calcitic, aragonitic or mixed mineralogy, possibly depending on the composition of the sea water at their time of formation, as with inorganically precipitated grains like ooids (p.12). Calcareous worm tubes appear to have a two-layered shell: a thin inner layer of laminae arranged concentric to the tube, and a thicker outer layer of laminae inclined to the length of the tube. Vermiforms have a three-layered shell: an outer acicular layer, a central blocky prismatic layer, and an inner irregular microlamellar layer. It may be difficult to observe these features without an electron microscope.

175 shows a transverse section of a bundle of worm tubes from a Jurassic limestone. In this case the wall structure seems to have been replaced, suggesting that the tubes were originally aragonite. **176** shows the typical sections of numerous *Spirorbis* tubes.

175 Stained thin section, Middle Jurassic, Cotswolds, England, PPL, × 25.

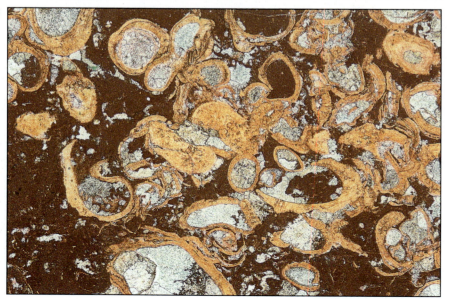

176 Stained thin section, Carboniferous, Northumbria, England, PPL, × 18.

Tentaculites

Tentaculites is a small conical shell of uncertain affinity which is abundant in some Devonian rocks. **177** shows two sections of *Tentaculites*: one elliptical tangential section (upper left) showing the characteristic ribs on the outside of the shell, and (to the right) a typical transverse circular section. The wall structure is finely foliated calcite, resembling the structure of some bryozoans and brachiopods, but in some cases, perforations in the shell wall can be seen at higher magnification. The other straight shells in the photograph are brachiopod valves.

Tintinnids

Tintinnids are widely distributed planktonic microorganisms, but only the few calcareous examples, also known as calpionellids, concern us here. The hard parts are vase- or cup-shaped, and thin sections show circular or horseshoe-shaped, thin-walled grains, rarely more than 100 m across. A number of these can be seen at high magnification in **178**. Tintinnids are best known from Late Jurassic and Early Cretaceous pelagic limestones from the Tethyan region.

177 Stained thin section, Devonian, Leon, Spain, PPL, × 24.

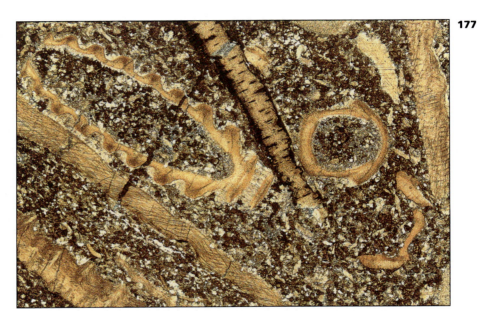

178 Unstained thin section, Upper Jurassic, Provence, France, PPL, × 200.

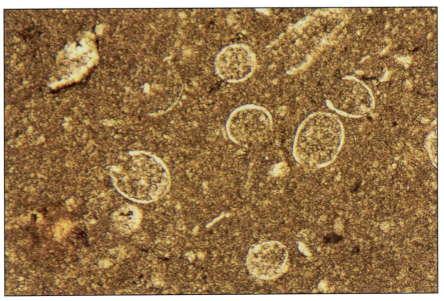

Carbonate Sediments and Rocks Under the Microscope

Radiolarians

Radiolarians are marine planktonic micro-organisms usually 100–200 m across. They are mostly siliceous, and spicules and spherical or conical tests may be preserved, especially in deep-sea cherts. In limestones, which are the concern of this book, they are rarely well-preserved and the original opaline silica has usually been replaced by calcite, obscuring the detail of the test. **179** shows a small circular cast now made of calcite in a carbonate mud matrix. Careful inspection shows that the margin appears 'toothed'. This is a radiolarian cast and the carbonate mud has partially infilled the pores on the surface of the test, giving the toothed appearance. Without this it would be difficult to be confident of the identification. **180** and **181** show a radiolarian-rich limestone in which the radiolaria are still partly composed of silica. Clear circular sections of radiolaria are evident in the photograph taken with plane-polarised light (**180**). In the view taken with polars crossed (**181**) it can be seen that some are composed entirely of fine quartz, showing first-order grey interference colours, but others are now entirely calcite or a mixture of calcite and quartz.

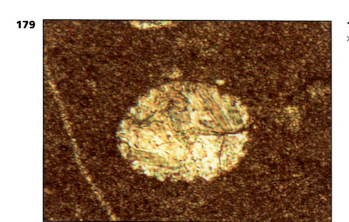

179 Stained thin section, Upper Cretaceous, Greece, PPL, × 200.

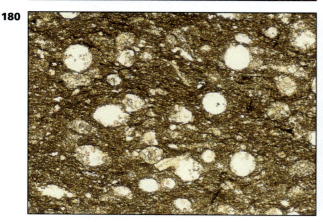

180, **181** Unstained thin section, Mesozoic, Greece, × 50, **180** PPL, **181** XPL.

Microbial structures

Micro-organisms, including bacteria and fungi, play an important part in the generation of fine-grained carbonate sediments. Some organisms become calcified and can thus be preserved as fosssils, but most only leave evidence of their presence in the overall structure of the rock. Structures believed to be formed by micro-organisms of this type are classed as microbial structures.

Calcified cyanobacteria, formerly classed as blue-green algae, have been given generic names although many different microbial organisms may have given rise to similar calcified structures. Small, simple tubes without partitions or branching are referred to *Girvanella*. These are seen as small 'bunches' in both longitudinal and transverse section in **182**. The more irregularly shaped clumps seen in **183** are referred to *Renalcis*, an important contributor to some Palaeozoic carbonate build-ups. Calcification of these organisms may depend on external factors, such as sea-water chemistry, rather than being controlled by the organism itself.

182 Stained thin section, Lower Carboniferous, Lancashire, England, PPL, × 65.

183 Stained thin section, Lower Carboniferous, Poland, PPL, × 55.

Carbonate Sediments and Rocks Under the Microscope

Cyanobacteria are the principal organisms involved in the construction of microbial mats (formerly known as 'algal' mats) and are therefore important in the production of the laminated carbonate sediments known as stromatolites. In the depositional environment these sediments consist of organic-dominated microbial mat layers and layers of sediment precipitated or deposited on the mat surface. As the sediment is buried, the organic matter decays and may leave a cavity which can be filled with internal sediment and/or cement, provided that it is not first destroyed by compaction. Ancient stromatolites rarely contain any organic matter, but comprise millimetre-scale laminations which are usually best recognized in hand specimen.

184 is a section through a young laminated sediment from below the surface of a modern tidal flat. Despite the absence of any organic matter, this sediment is interpreted as a stromatolite. It comprises alternations of fine peloidal sediment and carbonate mud. There are also layers such as that just above the centre which contain a lot of open space. These spaces, larger than grain-supported spaces, are known as fenestrae (p.158) and probably represent the original microbial mat layers which compacted on decay of the organic matter. 185 is a well-preserved ancient example of a stromatolite with a similar layering, comprising mud-rich layers and more sparry layers, the latter including cement filling the cavities left by the decaying microbial mat.

184 Unstained thin section impregnated with blue-dye-stained resin, Recent, Bahamas, PPL, × 21.

185 Stained thin section, Lower Carboniferous, Boulonnais, France, PPL, × 9.

DIAGENESIS

Diagenesis refers to all those processes which occur to a sediment after deposition, during burial and any subsequent uplift. Sediments which undergo deep burial or are involved in orogenesis, such that they experience high pressures and/or temperatures, will undergo metamorphism and no longer be classed as sediments. There is no hard and fast boundary between diagenesis and metamorphism. As a guide, limestones will retain sufficient of their primary features that the depositional texture is still recognisable, whereas their metamorphosed equivalents will show little or no sign of the depositional fabric. However, this rule cannot be applied to dolomites where replacement under sedimentary conditions can lead to total obliteration of the original texture. Diagenesis certainly embraces processes that can occur at up to several kilometres of burial and temperatures well above 100°C.

Carbonate rocks are particularly susceptible to diagenesis partly because carbonate minerals are more soluble in water than many other naturally occurring minerals and so are subject to dissolution and reprecipitation. This is enhanced because one common primary marine carbonate mineral, aragonite, is metastable under sedimentary conditions. Diagenesis can begin on the sea-floor and, indeed, grains can be re-worked and re-deposited after some very early diagenesis such as micritisation or marine cementation. Much diagenesis, including stabilisation of the mineralogy, occurs under the influence of meteoric waters which may displace the depositional marine fluids in the pores as a result of a relative sea-level fall. Diagenesis occurs most rapidly in near-surface environments where there is a vigorous water flow, but the importance of compaction increases with depth, and other processes, such as cementation and dolomitisation, may also occur, albeit more slowly than near the surface. **186** shows the nomenclature of some diagenetic environments.

The text is ordered by process, such that all the pictures devoted to cementation are together, followed by compaction, neomorphism and dolomitisation, although, as with other parts of the book, there is a degree of cross-referencing. It would be possible to order the book by diagenetic environment, but that would require a greater degree of interpretation than we believe appropriate to a book of this type.

Micritisation

Micritisation is the process whereby the margins of carbonate grains are replaced by micrite at or just below the sediment/water interface. The process involves microbes attacking the outside of grains by boring small holes in them, which are later filled with micrite cement. Skeletal particles are particularly prone to attack and in extreme cases micritisation can lead to complete alteration of the original grain and production of peloids (p.24). In some cases it can be difficult to distinguish micritisation from an external micrite coating which would strictly then make a coated grain (p.9), especially as both processes may affect the same grain. Where micritisation has led to complete circumgranular alteration, the micritic rind of the grain is called a micrite envelope. This should not be confused with micrite cements which may form an external coating around grains.

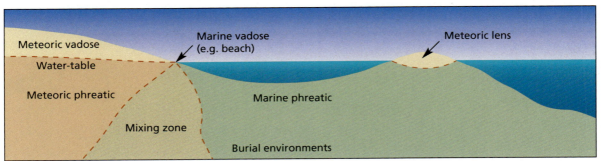

186 Sketch cross-section through a typical carbonate shelf showing the principal diagenetic environments.

Carbonate Sediments and Rocks Under the Microscope

Products of micritisation appear in numerous plates and are particularly well seen in 30, 58 and 74. 187 depicts a rock in which the depositional texture is only visible because of the presence of micrite envelopes around the original grains. The rock consists of molluscan casts, most of which are bivalves, although there is a transverse section of a gastropod near the centre of the picture. All were originally aragonite that has dissolved during diagenesis, only the micrite envelopes remaining to outline the shell shapes. A sparry calcite cement showing some compositional zoning has filled all intergranular and mouldic pore-spaces. It is likely that cementation and aragonite solution were closely linked in time, and that some intergranular cement was present before the shells dissolved. Otherwise it is unlikely that the micrite envelopes could have withstood compaction and retained the texture of the rock after solution of the aragonite.

Pedogenic features

A wide range of petrographic features can develop in limestones as a result of subaerial exposure and soil development, although explanation and illustration of the full range of features is beyond the scope of this book. Cementation and grain dissolution may occur in the soil zone and have been illustrated elsewhere in this book. Pisoids and other coated grains are characteristic and are also figured under their respective headings.

Structures related to root growth and decay are important in soils, including rhizocretions, which are coated or partially replaced roots. Sections of two rhizocretions can be seen just above the centre of 188. They consist of laminated micritic coatings around areas of sparry calcite (clear), the space originally occupied by a decaying root. Rhizocretions are often associated with alveolar septal fabric which comprises thin walls of micritic calcite separated by pore-spaces, as seen between the coralline algae (top) and coral fragment (right of centre) in 189. At high magnification the alveolar septal fabric is seen to consist of an unusual form of calcite known as needle-fibre calcite, illustrated in 190, taken with polars crossed. In ancient fully cemented limestones it may be necessary to use an electron microscope to demonstrate the presence of needle-fibre calcite. Alveolar septal structure is interpreted as a microbial structure related to fungal growth within soils.

Another structure attributed to fungi and roots is *Microcodium*, seen in 191. It is sometimes described as 'corn on the cob-like' and consists of a spherical or cylindrical structure with a hollow centre surrounded by radiating prisms of calcite. In the photograph the centres are filled with clear calcite cement and the radiating prisms are brownish and inclusion-rich. The rhombic crystals are dolomite partly replaced by quartz.

187 Stained thin section, Lower Jurassic, South Wales, PPL, × 25.

Diagenesis

188 Stained thin section, Lower Carboniferous, Cumbria, England, PPL, × 28.

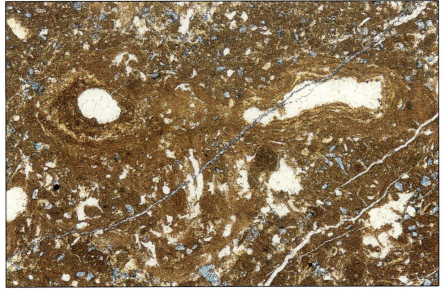

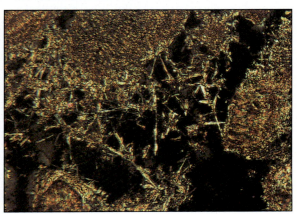

189, **190** Unstained thin section impregnated with blue-dye-stained resin, Quaternary, Barbados, **189** PPL, × 30, **190** XPL, × 140.

191 Unstained thin section impregnated with blue-dye-stained resin, Lower Permian, Spitsbergen, PPL, × 28.

103

In semi-arid flood-plain environments, calcium carbonate may be precipitated in the soil zone from waters drawn upwards by evaporation. The result is nodules of calcium carbonate, known as 'cornstones', which occasionally amalgamate to give a concretionary limestone bed in an otherwise terrigenous clastic succession. **192** is a section of cornstone showing coarsely crystalline calcite which is markedly twinned, supporting grains of detrital quartz. In order to produce a sediment which now appears calcite-cement-supported, the calcite must have grown displacively with respect to the original framework grains or has, in part, replaced them.

Cementation

Cementation is the process of precipitation of space-filling crystals. Most cements in carbonate sediments are themselves carbonates, although other minerals such as evaporites and quartz occur locally. Carbonates cements are precipitated in many different environments, from marine through meteoric to burial (**186**). Cements precipitated in marine environments form only a small proportion of the total cement in limestones, but are nevertheless locally important in reefs, on beaches, in some shallow marine grainstones and in deeper water muds. Marine cements may be aragonite or calcite and show a variety of morphologies. One might expect the mineralogy of carbonate cements to vary through time in the same way as the mineralogy of the precipitated grains, the ooids (p.12), relating to changing sea-water chemistry. However, although modern marine ooids are almost always aragonite, both aragonite and calcite cements are well known from Recent environments.

Recent marine cements commonly include acicular aragonite, bladed or prismatic Mg-calcite, micritic Mg-calcite and peloidal Mg-calcite. There are also some records of equant calcite cements from marine environments. This is a morphology normally associated with meteoric environments and it illustrates the care that must be taken in interpreting environments of cementation on the basis of crystal morphology alone.

Cements precipitated from near-surface meteoric waters are normally iron-free calcite with roughly equant crystal shapes. As a result of competitive growth, crystals usually increase in size away from substrates towards the centres of the original pore-spaces. This feature is called a drusy mosaic and characterises many limestones fully cemented in

meteoric environments. As sediment is buried, any oxygen in pore-waters is used up, and in the reducing burial environments calcite cements may contain ferrous iron (Fe^{2+}) substituting for some of the calcium. Waters move more sluggishly in the deep subsurface and crystals grow more slowly, but are often entirely coarse-grained and fabrics lack drusy mosaics.

Environments of cementation are also described with respect to the position of the water-table (**186**). In the vadose zone, above the water-table, pores are not filled with water, but with a mixture of water and air. Water collects on the undersides of grains and as menisci at grain contacts, so cement is unevenly distributed in the same fashion. Below the water-table, in the phreatic zone, pores are filled with water and cement grows more or less evenly on all surfaces. Circumgranular cements of equal thickness are said to be isopachous. This property is lost as the cement grows to fill all available pore-space.

193 shows an example of an acicular aragonite cement in the pores of a Recent grainstone sediment. Although the cement is broadly isopachous and with crystals orientated with their long axes normal to the grain surfaces on which they are growing, the rather ragged outline of the cement is characteristic. In this case pores are filled with blue-dye-stained resin to protect the cement during making of the thin section. Despite this, cement has been lost from the pore at the bottom of the photograph.

194 shows quite a complex cementation history despite coming from a relatively young tropical island beachrock. The grains at the top of the picture are coral fragments, and a transverse section of a gastropod is seen on the left. The first generation of cement is an isopachous crust of acicular aragonite and this is succeeded by a rather more irregular crust of dark-looking micrite cement. Just to the right of the centre of the photograph a few small peloids seem to be associated with this generation of cement. They are probably an internal sediment washed into pore-spaces during diagenesis. The third and final generation of cement is further acicular aragonite and some pore-spaces remain. Note that the first generation of acicular aragonite is absent in the intragranular porosity within the gastropod, where the micrite cement is more obvious. These cements are all marine, the changes reflecting varying conditions, perhaps as a result of changing sea-levels.

Diagenesis

192 Stained thin section, Devonian, South Wales, PPL, × 42.

193 Unstained thin section, impregnated with blue-dye-stained resin, Quaternary, Barbados, PPL, × 60.

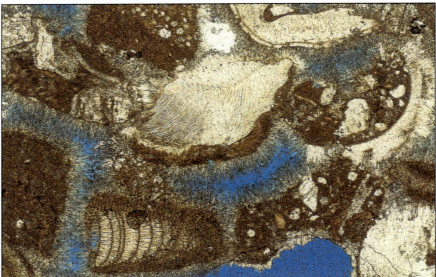

194 Stained thin section, Quaternary, Rodriguez Island, Indian Ocean, PPL, × 30.

Carbonate Sediments and Rocks Under the Microscope

195 shows quite a high-magnification view of a cemented Quaternary carbonate sand. The dominant grains are molluscan fragments showing a crossed-lamellar structure (p.36). A thin isopachous crust of small crystals can be seen on all grains, including the clear detrital quartz grain in the upper left part of the picture. This crust of crystals is very thin (about 10–20 m), but despite this the hand specimen from which this section was taken was a well-lithified rock. This is another marine-cemented rock, and the crystal morphology, although not clearly seen in the photograph, is of small prismatic crystals of calcite growing with their long axes perpendicular to the grains.

196 shows a lightly cemented mixture of sand-sized quartz and broken shell fragments. The cement is not present on all grain surfaces, but is mostly concentrated at grain contacts. This meniscus fabric is diagnostic of vadose environments. In this case, the cement is micritic calcite of probable marine origin and appears dark in the photograph. The pores are filled with blue-dye-stained resin.

195 Stained thin section, Quaternary, Rabat, Morocco, PPL, × 85.

196 Unstained thin section impregnated with blue-dye-stained resin, Quaternary, South Wales, PPL, × 40.

Diagenesis

A vadose fabric is also seen in **197** and **198**. In this case the sediment is an ooid grainstone with locally developed meniscus cement at grain contacts. The overall effect is to 'round off' the pore-spaces. The fabric is most clearly seen in the view taken with polars crossed (**198**), especially in the upper left and lower right parts of the photograph. In contrast to the cement in **196**, the cement shown here is made up of larger crystals, but with no preferred shape or orientation evident. This is characteristic of cement precipitated from meteoric waters. Despite this evidence of meteoric water diagenesis there is no evidence for the alteration of the primary aragonite ooids, at least at this magnification. **198** also shows examples of the pseudo-uniaxial extinction cross seen in most Recent aragonite ooids when viewed in thin section with crossed polars (p.10).

197, **198** Unstained thin section, Quaternary, Caribbean, × 35, **197** PPL, **198** XPL.

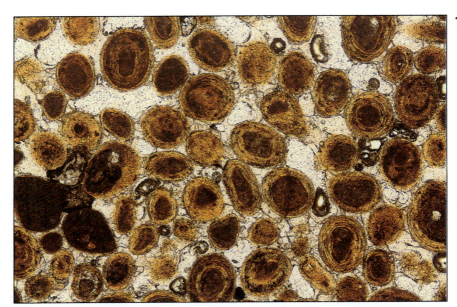

197

198

Carbonate Sediments and Rocks Under the Microscope

A further example of a vadose fabric is seen in **199** and **200**. Here, rather than a meniscus cement, a 'dripstone' fabric, in which cement is absent on the upper surfaces of grains, but thickens down the sides to the undersurfaces, is well seen. The dripstone fabric is developed in the first generation of relatively fine-grained cement which appears cloudy in contrast to the coarse clear cement which fills the pore-spaces. This is well seen in the low-magnification view (**199**). **200** is a higher-magnification view of the left centre part of **199** and is taken with polars crossed. It shows that the first-generation cloudy cement is crudely prismatic with crystals growing normal to the grain surface. This type of fabric occurs with marine cements, especially in beachrocks as in this case, but can also be found in cave deposits (speleothem). Where the cement appears cloudy in **199** and **200** this is caused by the abundance of fluid or solid inclusions within the crystals.

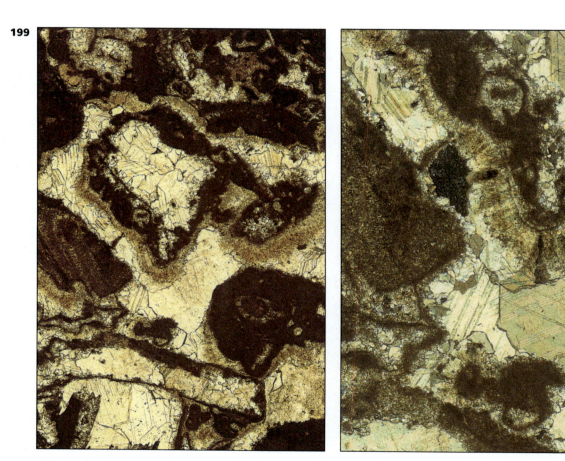

199, 200 Unstained thin section, Lower Cretaceous, Massif de Chartreuse, France, **199** PPL, × 20, **200** XPL, × 45.

Diagenesis

201 and **202** show three generations of cement. The first generation is a more or less isopachous crust of fine crystals, the second is a cloudy brown cement with a dripstone fabric, and the third is a clear anhedral spar which fills the remaining porespaces. The first-generation cement is most clearly seen where it separates the external surfaces of the grains from the second generation of cement (for example, on the almost circular grain just above and to the left of the centre of the field of view). At this magnification its fabric is not clearly seen. In fact it is finely prismatic with a radial fabric and is probably a marine precipitate. The radial fabric is indicated in the view taken with polars crossed, where the cement crystals of this generation are in extinction in N, S, E and W positions (parallel to the edges of the photograph). The second, brownish, inclusion-rich cement is clearly vadose in origin, and of coarsely prismatic dripstone morphology, but the composition of waters from which it was precipitated is uncertain. The third generation of cement is a meteoric phreatic or burial cement. The open, uncompacted texture of this sediment is characteristic of rocks that were substantially cemented early in diagenesis, before significant burial.

201, **202** Unstained thin section, Lower Carboniferous, Lancashire, England, × 35, **201** PPL, **202** XPL.

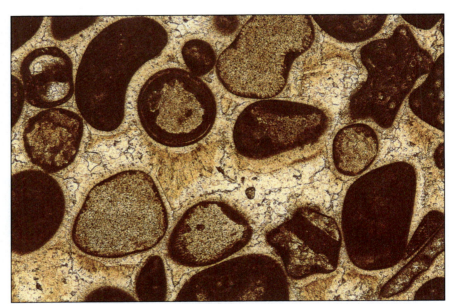

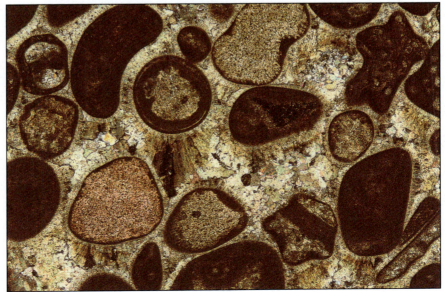

Carbonate Sediments and Rocks Under the Microscope

203 shows two sediments separated by an erosion surface running across the centre of the picture. The sediment below the erosion surface has at least two generations of cement, an early crust of yellowish crystals and a later clear anhedral spar which shows a drusy mosaic in the large pore to the right. These are respectively a marine cement and a meteoric phreatic cement. Between these two generations there is some micritic sediment, which, in places, shows a peloidal texture. This might be a cement, but could also be an internal sediment. The sediment above the erosion surface shows only a single well-developed cement generation of anhedral spar. The lower sediment was cemented in a marine environment and subsequently eroded, leaving a lithified surface, called a 'hardground', on the sea-floor. This was later buried by further sedimentation and eventually the whole sediment was cemented under the influence of meteoric waters.

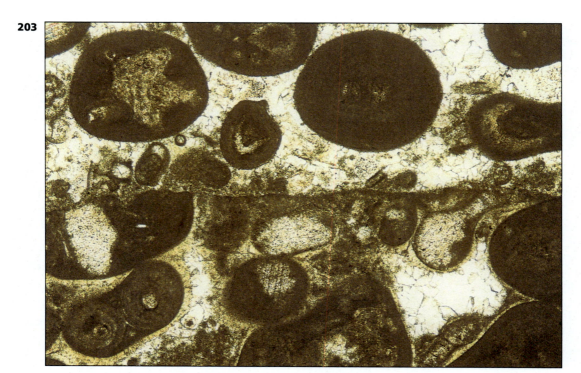

203 Unstained thin section, Lower Carboniferous, South Wales, PPL, × 40.

Diagenesis

Two generations of cement are clearly visible in **204** and **205**. In addition to the difference in crystal morphology between the two cement generations, staining shows a compositional difference. The first-generation cement is an isopachous radial-fibrous non-ferroan calcite and is likely to have been a marine precipitate. Since the fabric is well preserved, this is could have been a primary calcite cement. Although most radial-fibrous cements can be safely interpreted as marine in origin, it is rarely possible to be certain of their original mineralogy on the basis of thin-section petrology alone. Primary aragonite needle cements would not survive subsequent diagenesis and would be subject to solution or recrystallisation to calcite. In the latter case, it would be expected that elements of the original radial-fibrous fabric would be retained. The second generation of coarser, mauvey-blue stained cement is ferroan, and therefore was precipitated in reducing conditions, during meteoric phreatic or burial diagenesis. The sediment is a bioclastic grainstone containing molluscan casts and also lithoclasts (like the large one in the upper right part of the photograph), which were largely dolomite, although they are now dedolomite (p.147).

204, **205** Stained thin section, Triassic, Germany, × 30, **204** PPL, **205** XPL.

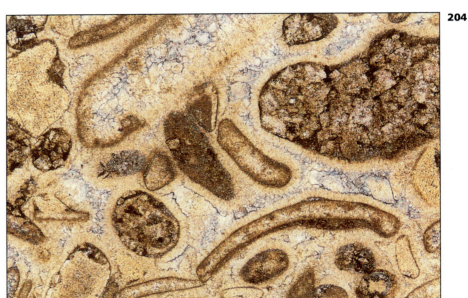

204

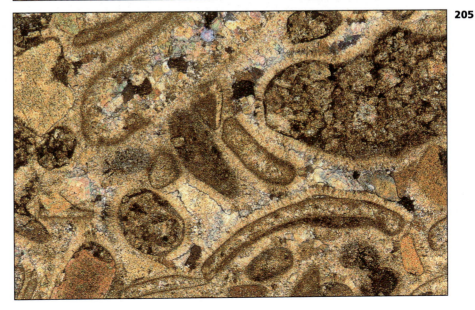

205

Carbonate Sediments and Rocks Under the Microscope

A special type of cement almost unknown from Recent environments, but common in many ancient sediments, especially in cavities in mounds and reefs of the Mid and Late Palaeozoic, is also thought to be a marine precipitate. This is radiaxial calcite or radiaxial fibrous mosaic. It consists of thick growths of fibrous calcite. Individual crystals are poorly defined, have undulose extinction and twin planes that are curved. The distinguishing feature of radiaxial calcite is that within crystals, the fast vibration directions (optic axes) converge away from the substrate (**206**). A variety of calcite with similar overall appearance, but with divergent optic axes away from the substrate is known as fascicular optic calcite. Radiaxial calcite is illustrated in **207** and **208**, where it is the main cement. Evidence for undulose extinction can be seen in the view with polars crossed (**208**), especially in the lower right part of the photograph where only the left side of each crystal is at extinction. Hemispherical growths of radiaxial calcite cement can be seen in the view taken with plane-polarised light, particularly just above the centre of the picture. This would have given a botryoidal appearance to the cement during growth. Modern botryoidal aragonite cements are well known, and perhaps radiaxial calcite is an ancient calcite 'equivalent'.

The possibility that some peloidal fabrics are actually a form a marine cement, precipitated with some microbial involvement, has already been mentioned (p.27). **209** is from a shell bed in which originally aragonite bivalves have been neomorphosed to calcite (p.128). The early cement is micritic but rather than occurring as an isopachous crust as in **194** or at grain contacts as in **196**, the micrite cement occurs as 'clumps' with a rounded outline, resulting in a vaguely peloidal structure.

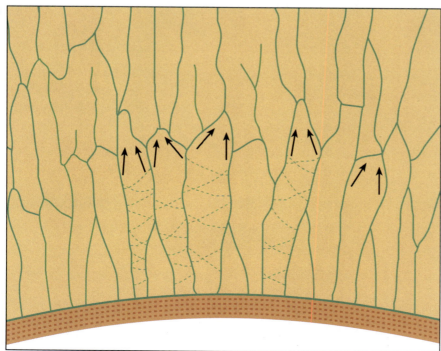

206 A sketch to show the typical fabric of radiaxial fibrous calcite. The crystals are growing away from a curved substrate (stippled). The arrows indicate the convergent optic axes within individual crystals. The effect of this is that as the microscope stage is rotated clockwise, the left-hand side of the crystal goes into extinction first and the shadow moves through the crystal from left to right. The curved twin planes shown on some of the crystals (broken lines) are not present in all examples of radiaxial calcite

Diagenesis

207, **208** Stained thin section, Lower Carboniferous, Derbyshire, England, × 20, **207** PPL, **208** XPL.

209 Stained thin section, Upper Triassic, England, PPL, × 45.

Carbonate Sediments and Rocks Under the Microscope

210 shows a very open fabric in a rock retaining high porosity, here filled with blue dye-stained resin. Two generations of cement are visible, both forming crudely isopachous layers. The initial, cloudy, inclusion-rich cement in which a radial fabric is just discernible is of marine origin, and the second, composed of clear 'blocky' or equant crystals is a typical meteoric phreatic cement. **211** and **212** show a grainstone in which the only cement is equant calcite of meteoric phreatic origin. This has not filled the pore-spaces, the remaining porosity being black in the view taken with polars crossed. The lack of evidence for significant compaction in this sediment is evidence that it was cemented fairly early in diagenesis, before any appreciable burial. In **213** a single generation of meteoric phreatic cement has led to the development of a drusy mosaic filling all the original pore-spaces.

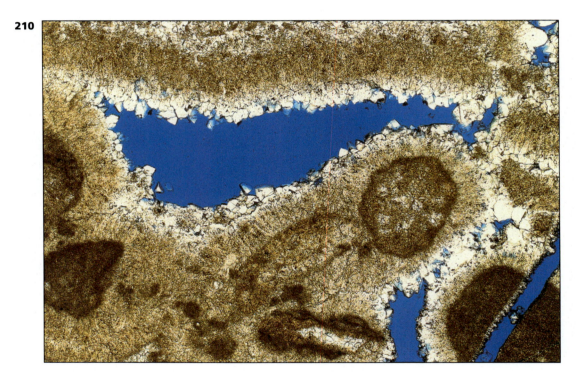

210 Unstained thin section, impregnated with blue-dye-stained resin, Upper Carboniferous, Spitsbergen, PPL, × 45.

Diagenesis

211, **212** Stained thin section, Upper Jurassic, Dorset, England, × 70, **211** PPL, **212** XPL.

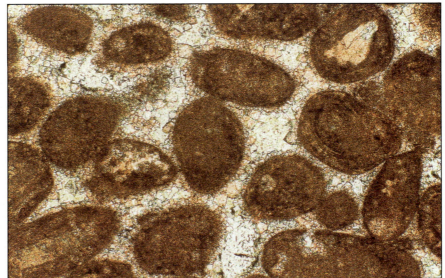

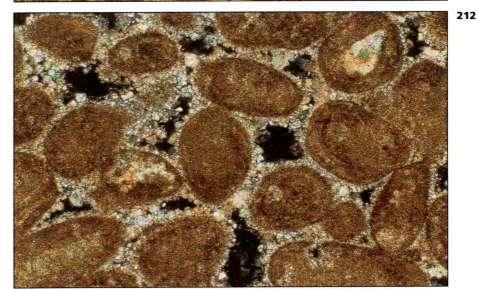

213 Unstained thin section, Middle Jurassic, England, PPL, × 42.

214 shows a fully cemented rock, with two cement generations. The first cement, occurring as thin isopachous crusts of radial crystals, is marine in origin and pores are filled with blocky crystals of probable meteoric phreatic origin. Notice that the molluscan casts, seen for example in the lower part of the photograph, have both cement generations on the outside, but only the second generation inside. Marine cementation took place while these grains were still aragonite and subsequently meteoric waters dissolved the aragonite molluscs and precipitated calcite cement in both the secondary mouldic pores and remaining primary intergranular pores.

In **215** an isopachous cement is well seen and at first sight this resembles marine cements illustrated here (e.g. **195, 203, 204**). However, on close inspection this cement does not have a distinct radial fibrous fabric and the crystals are more or less equidimensional. This cement is therefore likely to be a meteoric phreatic precipitate.

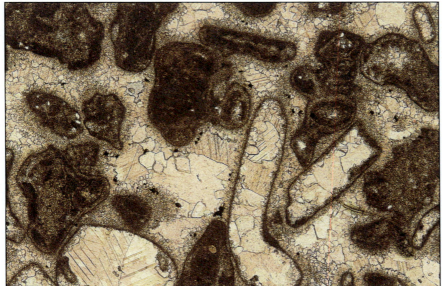

214 Stained thin section, Triassic, Oman Mountains, PPL, × 30.

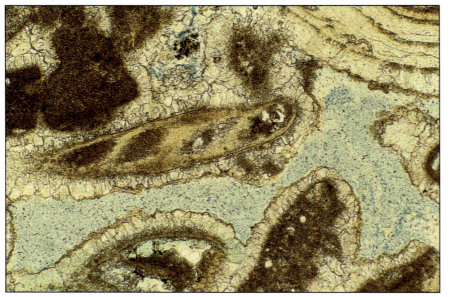

215 Unstained thin section impregnated with blue-dye-stained resin, Quaternary, Barbados, PPL, × 100.

Diagenesis

Although most early cements in limestones are non-ferroan, this is not always the case. **216** shows an oolitic grainstone cemented entirely by blue-stained ferroan calcite. The open, uncompacted texture suggests that cementation began early during diagenesis, before any compaction. The clear grains in this section are quartz. **217** shows another sample from the same unit as **216**. In this case, a ferroan calcite cement filling a shell mould (outlined by a micrite envelope) exhibits a drusy mosaic typical of meteoric phreatic conditions. There are a number of possible interpretations for these iron-rich cements. These limestones are part of a succession rich in fine-grained terrigenous clastic sediments. One possibility is that oxygen in the pore-waters was used up by decaying organic matter in the muds, such that pore-waters rapidly became anoxic, permitting the precipitation of ferroan cements in near-surface environments.

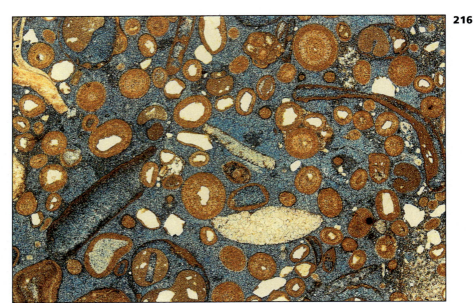

216 Stained thin section, Upper Jurassic, Dorset, England, PPL, × 25.

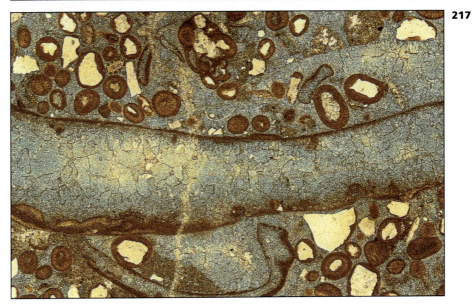

217 Stained thin section, Upper Jurassic, Dorset, England, PPL, × 25.

Carbonate Sediments and Rocks Under the Microscope

Syntaxial overgrowth or syntaxial rim cements, in which the cement crystals have grown by the extension of the lattice in depositional grains, are particularly obvious in echinoderm-rich rocks, where they are often the dominant cement. This is because echinoderm fragments themselves are large single crystal plates of calcite (p.76), so the syntaxial relationship is easily demonstrated. Furthermore, large single crystal hosts are a preferred site for cement precipitation, such that in a particular rock, syntaxial cements on echinoderms often develop at the expense of cement on other substrates. Syntaxial overgrowths are not diagnostic of a particular environment of formation; indeed they are often chemically zoned and may record precipation over a long period of time through successive environments. **218** and **219** are photographs of a typical crinoidal limestone from the Palaeozoic with a coarsely crystalline cement. The crinoids are cloudy in appearance and the cement is clear, and in the view taken with polars crossed (**219**), it can be seen that the cement crystals are showing the same interference colours as the adjacent echinoderms and are parts of the same crystal. The cements are, therefore, syntaxial overgrowths.

218, 219 Stained thin section, Lower Carboniferous, Lancashire, England, × 15, **218** PPL, **219** XPL.

Diagenesis

220 and **221** show single syntaxial overgrowths on echinoid spines. In **220** the area to the left and below the spine is the syntaxial overgrowth. One problem is how the space now occupied by the cement was formed, since it seems larger than the normal intergranular pore-spaces. It is known that in solutions which are just saturated, small crystals may dissolve and larger crystals grow at their expense. In this case, perhaps fine-grained carbonate in the immediate vicinity dissolved as the syntaxial overgrowth grew. In **221** the overgrowth has grown to include some of the original depositional grains in the sediment; this is described as a poikilitopic fabric. As in **220**, there is a very open fabric around the echinoid spine which, in this case, contrasts with the compacted fabric in the rest of the sediment. This suggests early initiation of overgrowth formation before significant burial and also, possibly, some solution of micritic material.

220 Unstained thin section, Upper Jurassic, Western High Atlas, Morocco, XPL, × 56.

221 Stained thin section, Middle Jurassic, England, PPL, × 40.

Carbonate Sediments and Rocks Under the Microscope

Syntaxial overgrowths are often present on substrates other than echinoderm fragments, although they are not as evident and are rarely developed at the expense of other cements. **222** and **223** show a rounded prismatic bivalve fragment (mostly pink-stained) with a blue-stained ferroan calcite cement. The edge of the grain is picked out by a thin micrite envelope. The view taken with polars crossed shows that some of the cement crystals adjacent to the bivalve fragment have the same extinction as the prisms that make up the shell and are therefore syntaxial overgrowths.

222, **223** Stained thin section, Middle Jurassic, England, × 36, **222** PPL, **223** XPL.

Diagenesis

Burial cements are often coarse grained and occasionally show poikilotopic textures. They are usually ferroan and can be shown to post-date compaction in many limestones. The cement in **224** is mauve-stained and thus somewhat ferroan, coarse-grained and post-dates compaction of the ooids. Grains can be seen to be squashed together (for example, right of centre in the photograph). In **225** there is a thin rind of early cement which has helped the sediment withstand compaction, and a coarse very pale blue-stained ferroan calcite cement filling the pores. Notice that the fractured bivalve fragment rimmed with micrite (top centre) is 'healed' by this burial cement. Burial cements are also figured in the following section dealing with compaction fabrics.

Dolomite, silica, pyrite and evaporite cements also occur in carbonate rocks. These are illustrated under the respective mineral.

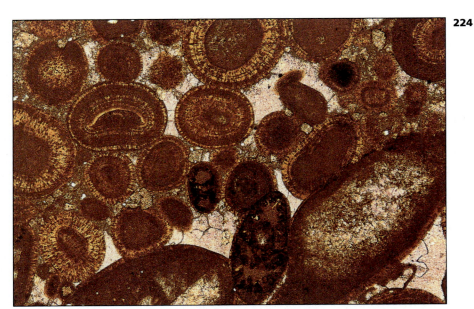

224 Stained thin section, Upper Jurassic, Western High Atlas, Morocco, PPL, ×35.

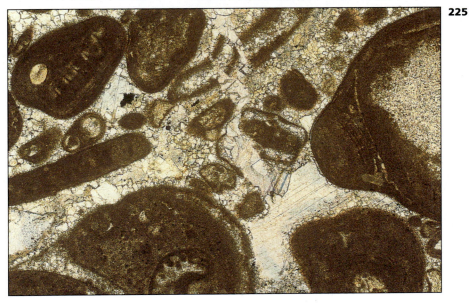

225 Stained thin section, Lower Carboniferous, South Wales, PPL, ×42.

Compaction and Tectonic Features

Many carbonates lose some of their original porosity as a result of compaction during burial. The effects of compaction are most pronounced in rocks in which early cements, such as marine and near-surface meteoric cements, are poorly-developed or absent. A well-sorted, rounded grainstone will generally have a primary intergranular porosity of 30–40% depending on the packing, and more poorly sorted and less rounded grainstones will have lower initial porosities. By estimating volumes of cement and remaining porosity, it is possible to establish the amount of porosity lost by compaction.

226 and **227** show an ooid grainstone which at first sight shows little evidence for compaction. However it is quite well-sorted and on deposition would have had an intergranular porosity of about 35%. The volumes of cement and remaining porosity can be most easily estimated from the view taken with polars crossed (**227**), where it can be seen that there is, in fact, little cement, seen as very thin fringes on the ooid surfaces. The porosity in the rock (black in **227**) is now less than 10%, so more than 20% of the original volume has been lost by compaction. Much of this will have been lost by re-packing of the fabric and by some 'squashing' of grains together. It can be seen that some original point contacts between grains have been modified, so that some adjacent ooids now meet in line contacts.

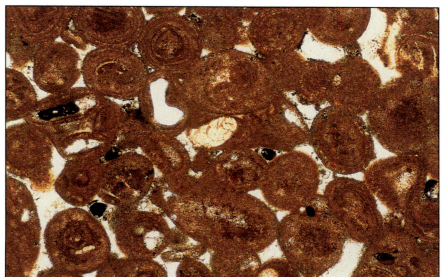

226, **227** Stained thin section, Upper Jurassic, Dorset, England, × 47, **226** PPL, **227** XPL.

In **228** compaction has gone further and ooids have been dissolved at their contacts with adjacent grains such that ooids now penetrate one another. This phenomenon is known as grain-to-grain pressure solution. The calcite for some late cements is supplied from this source. A pale mauve-stained, post-compaction ferroan calcite burial cement fills the remaining pore-spaces. In **229** there is a little dark micritic matrix visible, especially towards the top and bottom of the photograph. However, there is little matrix visible in the centre of the picture where there has been significant pressure solution. Grains have been dissolved along irregular lines known as stylolites. A stylolite is also illustrated in **254**.

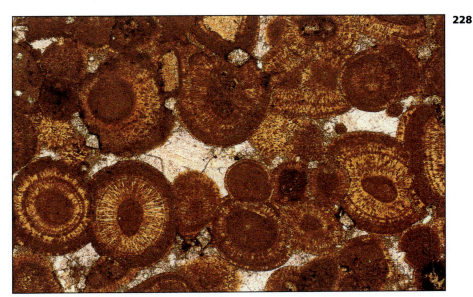

228 Stained thin section, Upper Jurassic, Western High Atlas, Morocco, PPL, × 42.

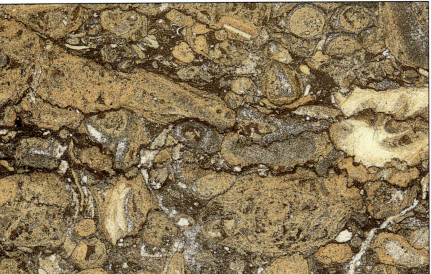

229 Stained thin section, Middle Jurassic, Western High Atlas, Morocco, PPL, × 15.

Carbonate Sediments and Rocks Under the Microscope

230 shows a bioclastic limestone comprising echinoderm and brachiopod fragments from a sequence of muddy carbonates. However, any matrix or porosity in the original sediment has been totally lost during compaction and a welded mass of bioclasts remains. In this sample there is little evidence for mechanical fracturing, and compaction involved re-packing of grains and pressure solution.

231–234 show the effects of compaction on individual grains. **231** is a bioclastic wackestone in which the bioclasts are ostracods. Compaction has fractured the single ostracod valve, but has had a more dramatic effect on the complete two-valved shell, which has completely collapsed, probably because the inside was free of sediment or cement. This is a good example of mechanical compaction through grain fracture and breakage.

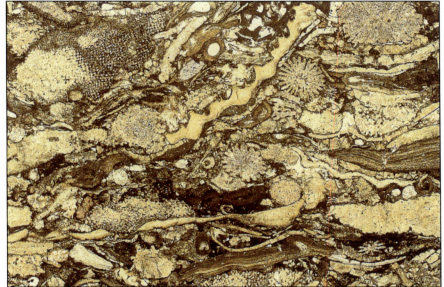

230 Unstained thin section, Lower Carboniferous, Lancashire, England, PPL, × 42.

231 Stained thin section, Miocene, Mallorca, Spain, PPL, × 100.

Diagenesis

232 is a sediment in which many of the grains are ooids or superficial ooids (p.14), mostly with shell-fragment nuclei. Just above the centre of the photograph there is an E–W elongated grain consisting of a fractured superficial ooid coating cemented by sparry calcite. This was formerly an aragonite mollusc fragment, coated with a thin calcitic oolitic crust. During diagenesis the aragonite dissolved, leaving a shell mould supported only by the thin oolitic coating. This collapsed and the sediment was later cemented by sparry calcite. The same feature is commonly seen with micrite envelopes around aragonitic grains. **233** is from an ooid grainstone with an early marine rim cement. The sediment was compacted and the outer margins of part of the uppermost ooid plus the rim cement have flaked or spalled off the grain. This can be seen in the centre of the field of view. Subsequently, calcite cement has filled the remaining pore-spaces.

232 Unstained thin section, Middle Jurassic, England, PPL, × 70.

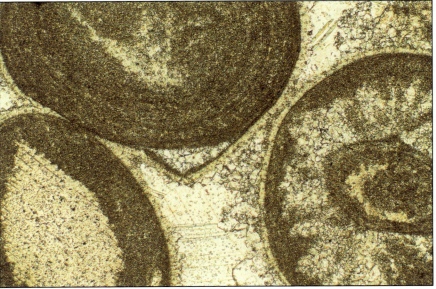

233 Unstained thin section, Lower Carboniferous, South Wales, PPL, × 100.

Carbonate Sediments and Rocks Under the Microscope

234 is a bioclastic wackestone with casts of gastropods. The original aragonite of the gastropods dissolved after the carbonate mud was sufficiently lithified to support the moulds so formed. In places, compaction affected the moulds such that the sediment filling the chambers has been pushed against the intergranular sediment, closing up the mould. This is best seen in the large shell which occupies much of the field of view. After some compaction a blue-stained ferroan calcite burial cement was precipitated in the remaining mouldic pores.

Very few limestones that have suffered any burial escape fracturing or veining, and in many older carbonates the density of veins is such that some appear in any low-power view of a peel or thin section. In **235** two generations of veins can be seen cutting a carbonate mudstone. The two veins running NW–SE are filled with non-ferroan calcite (stained pink) and these are cut by a later ferroan calcite vein (stained blue).

Carbonates in areas that have suffered significant folding and faulting may show signs of deformation. One feature which commonly develops is twinning of coarse calcite crystals such as cement or echinoderm fragments. This may occur simply as a result of burial, but in the case of **236** and **237** echinoderm fragments are not only twinned, but show undulose extinction (visible in the view taken with polars crossed, **237**), which has developed as a result of deformation.

Large crystals, such as echinoderm plates, are often the first grains in a limestone to show evidence of deformation. In **238** and **239** an echinoderm fragment, originally a single crystal, is now polycrystalline with individual crystals in a similar, but not identical, orientation. In **240** deformation has altered the shapes of the grains, elongating grains in the E–W direction of the photograph, as well as introducing twinning (stripy pattern) in echinoderm fragments between the larger coated grains (e.g. bottom left-hand corner of field of view).

234

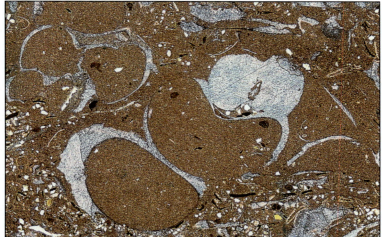

234 Stained thin section, Upper Jurassic, Dorset, England, PPL, × 20.

235

235 Stained thin section, Lower Carboniferous, Derbyshire, England, PPL, × 65.

Diagenesis

236, 237 Stained thin section, Lower Carboniferous, South Wales, × 25, **236** PPL, **237** XPL.

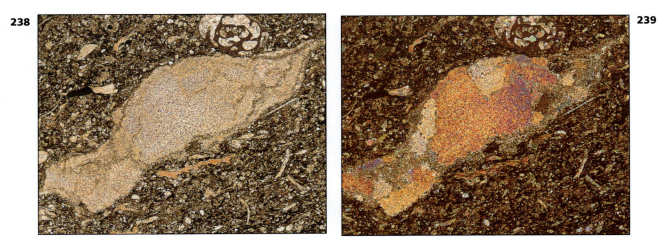

238, 239 Stained thin section, Lower Carboniferous, Lancashire, England, × 18, **238** PPL, **239** XPL.

240 Unstained thin section, Lower Carboniferous, South Wales, PPL, × 40.

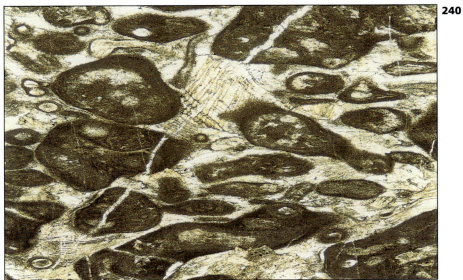

127

Neomorphism, Microspar and Pseudospar

The term neomorphism is used for processes of *in situ* replacement of one mineral by another of similar composition. In carbonate sedimentology, the term usually refers to aragonite to calcite transformations. Strictly, the term recrystallisation involves no mineralogical change and thus refers to modification of calcite or dolomite fabrics, although it is often used rather loosely since it is not always possible to demonstrate whether or not a mineralogical change has occurred. Most neomorphism and recrystallisation involves an increase in crystal size and is called aggrading, but there are some examples of crystal size reduction known as degrading. Another term in common use for in situ processes is calcitisation. This can be used for the neomorphism of aragonite to calcite, but is also used for dedolomitisation or replacement of evaporite minerals by calcite. The key point about all these processes is that solution and re-precipitation take place across a thin film and without large-scale solution and porosity formation. The precipitating material does not, therefore, show typical cement fabrics such as drusy mosaics and there are often relicts of the original fabric retained in the new crystals.

Neomorphism can most easily be demonstrated where formerly aragonite bioclasts or ooids have been calcitised, retaining elements of the original structure. It is important to distinguish this mode of preservation from complete solution of fragments leading to mould formation, followed by infill with cement to form a classic cast. Calcitised bioclasts have already been figured under bivalves (**59**), gastropods (**67**) and corals (**92, 93**), and calcitised ooids are illustrated in **16** and **17**. **241–243** show further examples of calcitised bivalves. In the low-magnification view (**241**), the shell fragments can be seen to comprise brownish, inclusion-rich calcite, with, in places, a visible laminated structure.

241 Unstained thin section, Upper Triassic, England, PPL, × 17.

Diagenesis

In **242** and **243**, which are higher-magnification views of the shells in the lower right-hand part of **241**, the coarsely crystalline nature of the calcite is evident. This type of neomorphic calcite often displays pseudo-pleochroism. Calcite in a thin section of standard thickness shows a marked change in relief on rotating the microscope stage, and when the crystals are inclusion-rich this phenomenon is more marked, such that there is the appearance of pleochroism from colourless to brown. This can be seen in these two views, **243** being taken with the polariser rotated through 90° with respect to **242**. This has the same effect as rotating the microscope stage through 90° when the section is observed with polarised light.

242, 243 Unstained thin section, Upper Triassic, England, PPL, × 65. In **243** the polariser has been rotated by 90° with respect to that in **242**.

Carbonate Sediments and Rocks Under the Microscope

Distinguishing the products of aggrading neomorphism or recrystallisation of carbonate mud matrix and fine cements from products of direct cementation or from primary sediments is often difficult. Criteria for distinguishing cement from neomorphic spar have been discussed by Folk (1965) and Bathurst (1975). Neomorphic fabrics usually comprise irregular crystals with curved and embayed boundaries, a variable crystal size with remnants of micritic sediment, and the presence of carbonate grains 'floating' in a spar matrix. Despite these criteria, many fabrics, particularly finer-grained ones, may remain of uncertain origin. The term microspar is used for neomorphic fabrics of 5–30 m average crystal size, and pseudospar for neomorphic fabrics of average grain size >30 m. The term micrite is used for all fabrics of crystal size <5 m.

244 and **245** show fabrics in which presumed original carbonate muds have suffered aggrading neomorphism. In **244** there is an equant 'matrix' of granular calcite which also seems to be replacing the margins of grains. The crystals average 40–50 m in size and are thus interpreted as pseudospar. **245** shows a matrix of varying grain size separating corroded dolomite rhombs. The matrix contains patches of micrite (darker red-brown stain) together with microspar and pseudospar. In the absence of any well-defined structure, the latter are interpreted as the products of aggrading neomorphism, with the micrite being remnants of the original sediment.

244 Unstained thin section, Lower Carboniferous, Cumbria, England, PPL, × 35.

245 Stained acetate peel, Lower Carboniferous, Derbyshire, England, PPL, × 42.

Diagenesis

Some microspar-sized fabrics are not the products of aggrading neomorphism. Cements of microspar size have been illustrated in **195**, for example, and other fabrics may be silt-sized sediments. This is particularly true of internal sediments, which are commonly described as 'crystal silts'. **246** shows fenestral porosity (p.156) in a carbonate mudstone, the lower part of which is filled with sediment and the upper part with cement. Such partial fills of sediment within cavities indicate the horizontal at time of deposition and are called geopetal infills. In this case it can be seen that the geopetal sediment is coarser than the original rock-forming carbonate mud. The geopetal sediment is of microspar size, whereas the host sediment is micrite. **247** is from a laminated fine-grained sediment in which thin silt-sized carbonate layers alternate with thicker finer-grained layers. Each fine-grained layer, however, is of microspar-size at the base and grades up into micrite. This grading, visible in the photograph and resulting in an upwards-darkening of colour within each layer, could not develop in such a systematic way by neomorphism, but must be a primary feature of the sediment.

Degrading recrystallisation is known from echinoderm limestones that have been buried and perhaps subject to deformation. An echinoderm fragment showing this reduction in crystal size is illustrated in **238** and **239** (p.127).

246 Unstained thin section, Upper Jurassic, Provence, France, PPL, × 40.

247 Stained thin section, Lower Carboniferous, Derbyshire, England, PPL, × 30.

Dolomites

Many carbonate sediments contain the mineral dolomite, $CaMg(CO_3)_2$, and some are totally made of dolomite. Dolomite is largely a secondary, replacive mineral although it can also occur as a cement. However, it can form at many different stages of diagenesis, from soon after deposition to deep burial, and from waters of many different compositions, from relatively dilute mixed marine and meteoric waters through sea water to hypersaline and burial brines. Manganese and particularly iron may substitute for the magnesium in dolomite, and iron-rich varieties are known as ankerite, $Ca(Mg,Fe)(CO_3)_2$. Although petrographic studies are of vital importance in studying dolomites, in very few cases is it possible to be certain of the mechanism of dolomitisation without resorting to chemical analysis.

Dolomite recognition

Unlike calcite, dolomite usually occurs as euhedral rhombohedra which can often be seen to be replacive. Otherwise its optical properties are similar to those of calcite. Where a sediment has been totally dolomitised the rhombic shape of the dolomite crystals in thin section may no longer be apparent. An additional complication is that some dolomite crystals may be partially or totally replaced by calcite during later diagenesis (dedolomitisation, p.147). Many carbonate thin sections are routinely etched and stained (p.6) partly to aid the distinction between dolomite and calcite.

Partial dolomitisation

In partially dolomitised rocks, important features to look for include the size of the dolomite crystals and whether they are replacing particular components of the original sediment, in which case the dolomite is described as fabric selective, or whether they are replacing all components of the sediment in an indiscriminate fashion, in which case the dolomite is non-fabric selective. **248–250** show a series of carbonate mudstones with an increasing degree of dolomitisation. **248** shows about 10% dolomite in the form of small clear rhombs. A few triangular sections can also be seen, where rhombohedra have been sectioned across a corner. This is a stained section and the calcite mud is a reddish-brown colour. In **249** about 50% of the carbonate mud (dark) has been replaced by dolomite. The crystals are larger than those in **248** (note the different magnifications) and have cloudy centres and clear rims – a commonly observed feature of dolomites. In many cases the inclusions are of calcite, but they can also be fluid inclusions or inclusions of non-carbonate material, such as clay present in the original sediment. **250** shows a sediment that comprises dolomite (about 90%) together with remnants of almost opaque calcite mud. The dolomite crystals are inclusion-rich, giving them a cloudy appearance. The rhombic shape is visible where there are isolated crystals and at the margins of the dolomite areas, but is not evident where the crystals have grown together.

Diagenesis

248 Stained thin section, Upper Jurassic, Western High Atlas, Morocco, PPL, × 70.

249 Unstained thin section, age and locality unknown, PPL, × 35.

250 Unstained thin section, Upper Jurassic, Provence, France, PPL, × 22.

Carbonate Sediments and Rocks Under the Microscope

251–256 show cases where partial dolomitisation has affected grainy carbonate sediments as opposed to carbonate mudstones. 251 is a limestone in which the prominent grains are large micritic lumps and they are supported by a carbonate mud matrix. Clear dolomite rhombs are selectively replacing the matrix, although in places they 'eat' into the edges of the grains.

252 shows another example of fabric-selective dolomitisation where the intergranular material has been replaced in preference to the grains. The sediment was a coarse intraclastic grainstone and staining shows up the original calcitic components in shades of pink, red and brown with the dolomite unstained, clear or cloudy. The variation in degree of cloudiness has two causes. To a degree it results from the inclusion density in the crystals, but it is also dependent on the orientation of the crystals in the section. Because its two refractive indices are so different, dolomite shows variable relief in thin sections, and those crystals in an orientation such that they show higher relief are of a cloudier appearance than those crystals in an orientation which shows low relief. In the enlarged view of the centre part of this rock (253), it can clearly be seen that although the dolomite is preferentially replacing the intergranular areas, it has also begun to replace the margins of the grains. Some dolomite crystals are common to areas occupied both by original grains and former intergranular areas perhaps once occupied by calcite cement. In these cases the part of the crystal relacing the former grain is inclusion-rich and cloudier than the part of the crystal occupying the former intergranular area. Were dolomitisation to have proceeded to completion in the same way, the original texture of the rock would still have been apparent from the inclusion pattern. This is a form of mimicking dolomitisation (p.136). There are a few isolated rhombs of dolomite replacing the inner part of the grains (e.g. lower right).

Late diagenetic dolomitising fluids often use pathways such as fractures and stylolites to move through otherwise cemented rock. 254 is a section of a bioclastic, intraclastic limestone cut by a stylolite. Brown iron oxide minerals are concentrated along the stylolite, together with a few clear rhombs of dolomite.

251 Unstained thin section, Jurassic, Greece, PPL, × 31.

Diagenesis

252, **253** Stained thin section, Lower Carboniferous, South Wales, PPL, **252** × 20, **253** × 42.

254 Unstained thin section, age and locality unknown, PPL, × 36.

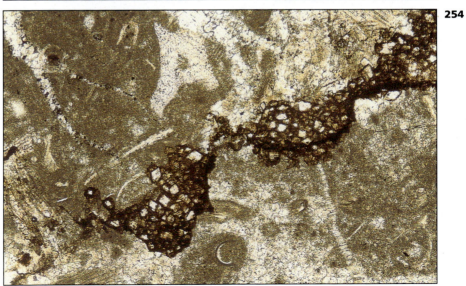

Carbonate Sediments and Rocks Under the Microscope

255 is a bioclastic and peloidal grainstone affected by a degree of dolomitisation. The two dolomite crystals in the centre show non-fabric-selective replacement, affecting both the brachiopod shell fragment and the intergranular cement, although, as in 252 and 253, the margin of the original grain is still visible within the dolomite crystals. 256 shows an example of a sediment that is 50% dolomite in which coarse dolomite rhombs indiscriminately replace both carbonate mud and sparry material. The dolomite crystals are quite cloudy due to the presence of small inclusions, and a few small red-stained areas of calcite are visible within some of the crystals. These may be the beginnings of dedolomitisation (p.147).

Total dolomitisation

Rocks which are totally dolomitised, where no remnant calcite from the original limestone remains, are dolomite rocks. Because of potential confusion over using the name 'dolomite' for both a mineral and a rock, some geologists use the term 'dolostone' for the rock, although it is not adopted here. Some dolomites show preservation of the original fabric of the rock, despite there being no remaining calcite; these are sometimes known as mimicking or mimetic dolomites. The fabric of the rock is usually preserved because the crystal size of the replacement dolomite mirrors that of the original sediment, or because of different inclusion densities in the dolomite crystals.

255 Stained thin section, Lower Carboniferous, Lancashire, England, PPL, × 42.

256 Stained thin section, Lower Carboniferous, Derbyshire, England, PPL, × 50.

Diagenesis

Occasionally, fabric preservation of the original limestone is so good that without staining or chemical analysis it may be difficult to believe that they *are* dolomites. This is the case with **257** and **258**. In fact both these thin sections were subjected to the staining process, but since both are entirely dolomite, lacking significant iron substitution, they remained completely unchanged. **257** is a stromatolite (p.100); during dolomitisation the layering was retained, along with the fine-grained nature of the sediment and the pelleted layers. Laminoid fenestrae (p.158) now filled with clear coarse dolomite are also evident, particularly in the lower part of the photograph. **258** is a photograph of a flat pebble conglomerate, an intraclastic rock formed by the reworking of desiccated carbonate mud. The carbonate mud clasts have been replaced by fine-grained dolomite, the geopetal (originally silt-sized?) sediment by slightly coarser dolomite, and the remaining spaces are filled with clear coarse dolomite. This last may be a primary dolomite cement or the replacement of a calcite cement.

257 Stained thin section, Permian, Oman Mountains, PPL, × 20.

258 Stained thin section, Permian, Oman Mountains, PPL, × 20.

259 shows a degree of depositional texture preservation in a coarse oncoid and micro-oncoid bearing sediment. In this case, in addition to some grain size variations within the oncoids, haematite staining of the matrix has promoted the textural preservation. In **260** the depositional texture is less well preserved, but slight changes in grain size and inclusion density allow the original sediment to be identified as a peloidal or oolitic grainstone. A shell fragment is also evident in the lower left part of the field of view.

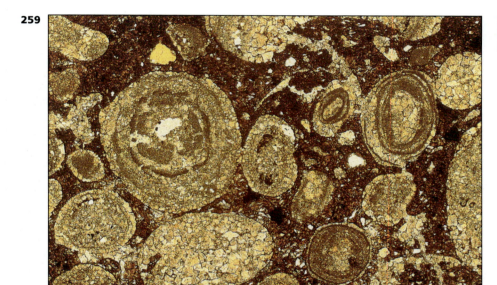

259 Unstained thin section, Lower Carboniferous, East Fife, Scotland, PPL, × 20.

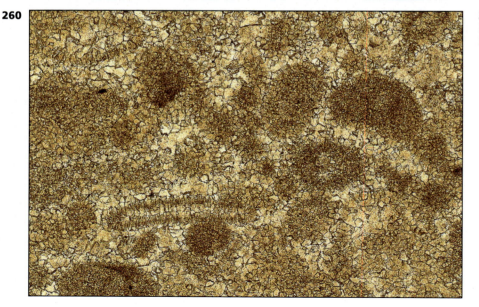

260 Stained thin section, Permian, Oman Mountains, PPL, × 35.

Diagenesis

261 is a photograph of a blue-dye-impregnated section of a porous dolomite. Although the original texture of much of the rock is not clear, fenestrate bryozoans are well preserved. In all these examples of mimicking dolomitisation, original micritic grains and matrix have been replaced by fine-grained dolomite, and pores or calcite spar by coarse dolomite. In **262**, however, there are rounded areas of coarse dolomite surrounded by a finer-grained matrix. The sediment has the appearance of having been an oolitic or peloidal wackestone. It would seem likely that the original grains had been dissolved and were moulds or casts at the time of dolomitisation.

261 Unstained thin section, impregnated with blue-dye-stained resin, Permian, Co. Durham, England PPL, × 32.

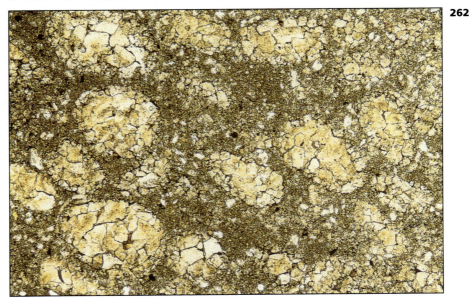

262 Unstained thin section, Middle Jurassic, Western High Atlas, Morocco, PPL, × 35.

263 is a medium-grained crystalline dolomite with apparently no sign of the texture of the precursor limestone. However, in such rocks where there is some variation in crystal size and inclusion density it is sometimes possible to get an indication of the original texture by inserting a sheet of plain white paper between the thin section and the stage. This has been done in this case and the result (**264**) shows that the original limestone was a grainy sediment, possibly oolitic or peloidal. A similar effect may sometimes be obtained by examining a polished section in reflected light.

263, 264 Unstained thin section, Permian, Oman Mountains, × 29, **263** PPL, **264** with white paper inserted between microscope stage and thin section.

Diagenesis

Many dolomites show little indication of the pre-dolomitisation texture of the rock. Nevertheless, it is important to describe the texture of the crystalline dolomite, including the crystal size and its variation (equicrystalline or inequicrystalline), and the crystal shapes (whether they are euhedral, subhedral or anhedral). 265 shows a coarsely crystalline dolomite with a fair proportion of straight boundaries. The fabric could be described as planar subhedral. In 266 there are fewer straight boundaries and the fabric is non-planar anhedral.

265 Unstained thin section, Permian, Oman Mountains, PPL, × 60.

266 Unstained thin section, Permian, Oman Mountains, PPL, × 60.

A planar fabric where most of the crystals are euhedral is usually only seen in dolomites which have some intercrystalline pore-space, such as the blue-dye-stained resin-impregnated section shown in **267**. In **268** and **269** intercrystal pore-space has been filled with a post-dolomitisation calcite cement, some of which appears to have grown syntaxially on the dolomite (e.g. the rhomb just below the centre). This is clearly seen in the view taken with polars crossed (**269**).

Dolomite crystals are often zoned; the characteristic inclusion-rich core and clear rim is particularly marked in **270**. In **271**, an example of a dolomite showing two crystal sizes, the larger crystals are multiply zoned.

Chemical zoning may sometimes be detected by staining. The coarse dolomite in **272** is probably filling a void in the finer dolomite. It consists of an inner, cloudy, unstained zone, lacking significant iron, and an outer, clear, turquoise-stained ferroan zone. The pinky-mauve crystal in the centre is calcite. Chemical zoning may also be revealed by cathodoluminescence (p.168).

267 Unstained thin section, impregnated with blue-dye-stained resin, Lower Carboniferous, Derbyshire, England, PPL, × 30.

268, **269** Stained thin section, Lower Carboniferous, Derbyshire, England, × 16, **268** PPL, **269** XPL.

Diagenesis

270 Unstained thin section, Upper Jurassic, Provence, France, PPL, × 55.

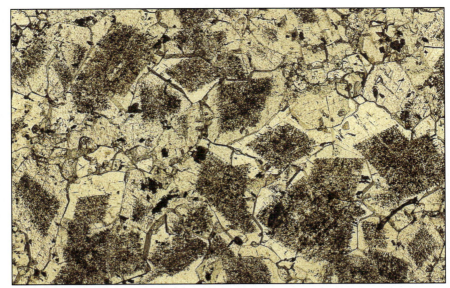

271 Unstained thin section, Cambrian, Senegal, PPL, × 13.

272 Stained thin section, age and locality unknown, PPL, × 15.

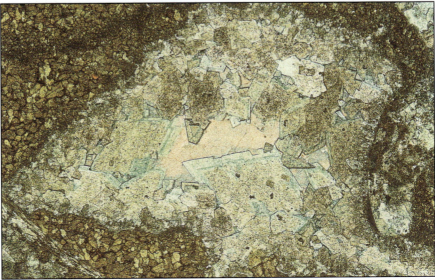

143

Carbonate Sediments and Rocks Under the Microscope

One form of dolomite which is thought to form at elevated temperatures during burial is baroque or saddle dolomite. This form of dolomite has a deformed crystal lattice and thus exhibits curved crystal faces, curved cleavages (where visible) and has undulose extinction. In **273** and **274** coarse baroque dolomite crystals have slightly curved faces, are inclusion-rich and some show sector-shaped subcrystals (e.g. the crystal near the left-hand side, just below the centre). The sweeping extinction, evidence for which is seen in **274**, is not the result of tectonic strain, but of lattice deformation during growth. Some intercrystal pore-spaces are empty, but others are filled with brownish-coloured clay which shows low-order interference colours in the view taken with polars crossed (**274**). Baroque dolomite crystals are often cements and are commonly ferroan.

273, 274 Unstained thin section, Lower Carboniferous, South Wales, × 32, **273** PPL, **274** XPL.

Diagenesis

275 and **276** show the cements filling a shelter pore (p.156) in a bioclastic grainstone. Twinned echinoderm fragments and a fenestrate bryozoan can be seen on the right-hand side of the picture. Initial cements are calcite, showing pale pink-stained non-ferroan calcite succeeded by strongly blue-stained ferroan calcite. These calcite cements are noticeably twinned (see, for example, top left). The final pore fill is a turquoise-stained ferroan baroque dolomite, with marked undulose extinction seen in the view taken with polars crossed (**276**). The contrast in stain colours of ferroan dolomite and ferroan calcite is well seen in this sample.

275, **276** Stained thin section, Lower Carboniferous, South Wales, × 30, **275** PPL, **276** XPL.

Carbonate Sediments and Rocks Under the Microscope

Baroque dolomite often fills veins and fractures and may be associated with sulphide mineralisation. **277** and **278** show baroque dolomite associated with opaque ore minerals. In **277** curved cleavages can be seen, and the extreme undulose extinction is apparent in **278**, taken with crossed polars, where only narrow zones in crystals are in extinction at once.

277, 278 Unstained thin section, Lower Carboniferous, North Pennines, England, × 15, **277** PPL, **278** XPL.

Dedolomites

Dolomites may be dissolved or replaced by calcite, particularly under the influence of meteoric water. Non-ideal (often calcium- or iron-rich) dolomites are particularly susceptible to alteration. The process is usually called dedolomitisation or, if dolomite is replaced by calcite, calcitisation of dolomite. 279 shows a single dolomite rhomb that is partially dedolomitised. Staining of the sample shows the distribution of calcite and dolomite very clearly. Because the replacement has occurred near the surface by oxidising meteoric waters, iron in the dolomite, which is in the reduced ferrous form, was oxidised during the replacement process and combined with oxide and hydroxyl ions in the water to produce a ferric iron precipitate rather than being incorporated in the calcite. Hence, as in this case, many dedolomites are heavily stained with brown iron oxides and hydroxides. In 280 an outer, probably iron-rich, zone of the dolomite has been preferentially dedolomitised. The resultant iron-stain masks the replacement calcite in many places.

279 Stained thin section, Lower Carboniferous, Derbyshire, England, PPL, × 150.

280 Stained thin section, Lower Carboniferous, North Yorkshire, England, PPL, × 55.

281 shows two former dolomite rhombs in a matrix of calcite mud. The two dolomite crystals have been replaced by a mosaic of fine calcite spar. In **282** former dolomite rhombs have been replaced entirely by calcite, with the exception of the crystal in the lower left-hand part of the photograph, which, for some reason, remains unaltered. Again the dedolomite is heavily stained with opaque iron oxide and in this case is surrounded by ferroan and non-ferroan calcite spar.

281 Unstained thin section, Upper Jurassic, Provence, France, PPL, × 60.

282 Stained thin section, Lower Carboniferous, Cumbria, England, PPL, × 32.

Diagenesis

Silica, Evaporite and Pyrite Cements and Replacement

Detrital quartz is common in many limestones and many carbonates also contain secondary silica, in the form of authigenic quartz grains, silicified shell fragments, chert nodules and occasionally silica cements.

Hexagonal sections of replacement authigenic quartz are seen in **283** and **284**, especially near the lower edge of the photograph. There is abundant silica in the rest of the rock, but since much of it is not euhedral it may be detrital in origin. Some of the authigenic quartz crystals show cloudy zones which are probably a mixture of quartz and carbonate, and indicate incomplete replacement of calcite by quartz. Two echinoderm fragments can be seen; the lower is twinned and has a syntaxial overgrowth cement (p.118).

283, **284** Unstained thin section, Lower Carboniferous, Lancashire, England, × 35, **283** PPL, **284** XPL.

Carbonate Sediments and Rocks Under the Microscope

285 and **286** illustrate silicified brachiopod shell fragments. The shells are of the type with a thick inner prismatic layer (p.52). Silicification is partial and the quartz is most easily distinguished from the calcite in the view taken with polars crossed (**286**) where the first-order interference colours contrast with the high-order colours of the carbonate.

285, **286** Stained thin section, Lower Carboniferous, Derbyshire, England, × 13, **285** PPL, **286** XPL.

Diagenesis

287 and **288** show an early quartz cement. It is most clearly seen lining the inside of the ostracod shell, but is also present elsewhere in the sediment. The relationships seen at the bottom of the photograph, where the valve is fractured, suggest that the quartz cementation predated compaction, since the cement does not line the fracture surface.

287, 288 Unstained thin section, Lower Carboniferous, Lancashire, England, × 35, **287** PPL, **288** XPL.

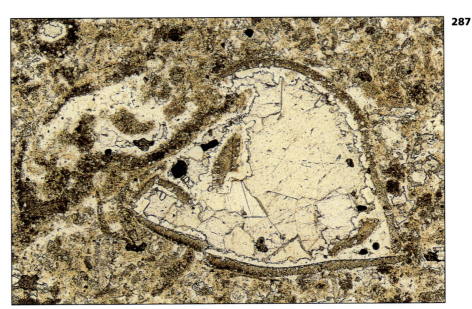

Carbonate Sediments and Rocks Under the Microscope

The variety of silica forming fibrous growths with a rounded botryoidal outline is chalcedony. This is seen as a cement in **289** and **290**, where it fills a fracture in a dolomite. There is also some (replacive?) quartz within the dolomite crystals close to the chalcedony. The quartz is transparent in the view taken with plane-polarised light (**289**), in contrast to the cloudy dolomite, but is best seen in the view taken with polars crossed (**290**) where the first-order interference colours of the quartz contrast with the high-order colours of the dolomite.

289, 290 Unstained thin section, Cambrian, Senegal, × 42, **289** PPL, **290** XPL.

Diagenesis

Anhydrite occurs as a replacement mineral and as a cement in carbonates, especially those associated with evaporites. Anhydrite forms rectangular crystals with cleavages at 90°, and shows bright interference colours up to mid third order. A large crystal of replacement anhydrite is shown in **291** and **292**, with further crystals along the right-hand edge of the photograph. Note that the boundaries of the anhydrite crystals clearly cross-cut the peloids. The anhydrite is replacing a peloidal grainstone with good intergranular porosity. Because they are relatively soluble, evaporites are often affected by further diagenesis.

291, 292 Stained thin section, Upper Jurassic, Saudi Arabia, × 17, **291** PPL, **292** XPL.

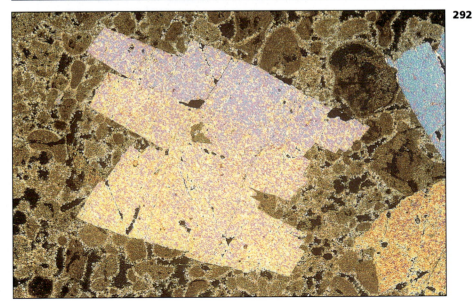

Carbonate Sediments and Rocks Under the Microscope

In **293** former anhydrite crystals in a carbonate mudstone, identified by their rectangular shape, have been replaced by mosaics of fine sparry calcite.

294 is a section from a shell bed with abundant pyrite cement, which appears black in the photograph. The shells are bivalves. There is abundant detrital quartz (clear) and a little glauconite (green). A pale-mauve-stained ferroan calcite cement fills the remaining pore-spaces.

293 Stained thin section, Upper Jurassic, Western High Atlas, Morocco, PPL, × 15.

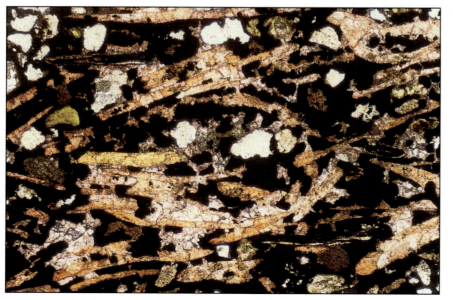

294 Stained thin section, Upper Triassic, England, PPL, × 47.

Diagenesis

Fluorite is occasionally found in carbonates adjacent to mineralised zones. Purple fluorite is well seen in **295** and **296** where it occurs in association with calcite and dolomite. The isotropic nature of the fluorite can be seen in the view taken with polars crossed (**296**). The rest of the isotropic material is orange-brown in colour in the plane-polarised light view (**295**) and is the mounting material occupying the pore-spaces. It would normally be colourless, but has been burned during exposure to an electron beam while observations of cathodoluminesence were being made (p.168). In this thin section the dolomite is very dark and cloudy as a result of dedolomitisation and precipitation of iron oxide (p.147) and the calcite is transparent.

295, **296** Unstained thin section, Lower Carboniferous, Derbyshire, England, × 15, **295** PPL, **296** XPL.

POROSITY

More than half the Earth's hydrocarbon reserves are contained within pore systems in limestones and dolomites, and therefore an assessment of the amount and type of any porosity in a carbonate sediment is an important part of any thin section description. Porosity can be described as primary, in which case it has been present in the rock since deposition, or secondary when it has developed during diagenesis. Porosity can also be described as fabric-selective if its location is controlled by particular parts of the depositional or post-depositional fabric of the rock. Porosity which is not fabric-selective typically cuts across the fabric of the rock. This division into fabric-selective and non-fabric-selective porosity types is the basis of the classification of carbonate porosity proposed by Choquette & Pray (1970), illustrated in **297**. As explained on p.8, many porous rocks are impregnated with blue-dye-stained resin before sectioning.

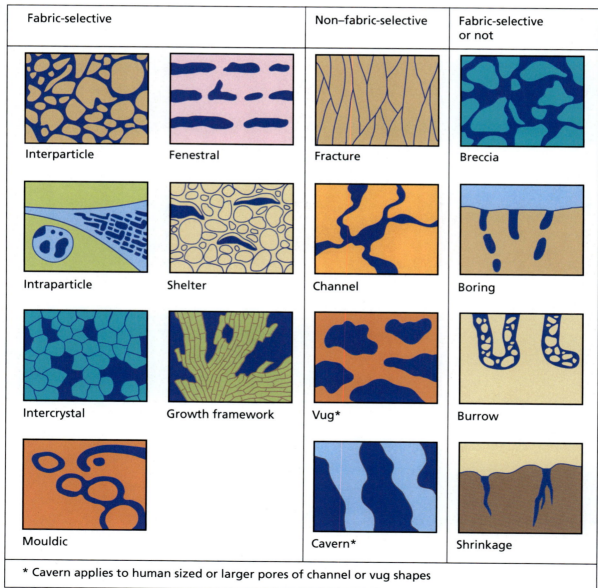

297

* Cavern applies to human sized or larger pores of channel or vug shapes

297 Classification of porosity in carbonate sediments according to Choquette and Pray (1970). Porosity is dark blue.

Porosity

298 is an example of a highly porous rock that would gladden the heart of any petroleum geologist. In fact, the porosity is remarkable since this sample comes from a depth of several thousand feet below the surface, where one might have expected compaction to have reduced porosity significantly. The rock is a bioclastic oolitic grainstone which received an early marine cement, seen as a thin isopachous crust of crystals around the depositional grains. The remainder of the primary pore-space between the grains remains unfilled so the rock can be said to have a high fabric-selective intergranular or interparticle porosity. There is a secondary mouldic porosity in this sediment, which is also fabric-selective since it formed by the solution of aragonite bioclasts. In the centre of the photograph there is a shell mould outlined by a micrite envelope, now slightly squashed, and elsewhere in the rock many of the rounded grains have dissolved centres (probably green algal fragments), for example to the left of the centre of the photograph. These types of porosity combine to give a total porosity of more than 20% of the rock volume. There has been some compaction, particularly evident where moulds have been fractured and squashed. The original point contacts between grains have also been modified to line or penetrative contacts by compaction. In the lower right part of the photograph, the porosity is somewhat less and there are some large clear cement crystals. These are syntaxial overgrowths on echinoderm fragments (p.118). This sample comes from the Jurassic Arab Formation, which contains some of the most important oil reserves in the Middle East and in which is found the world's largest known oil reservoir.

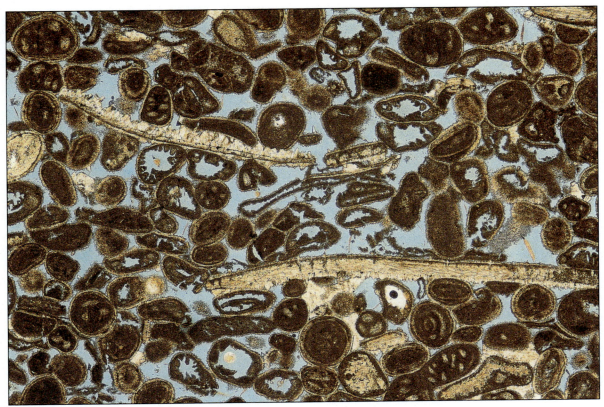

298 Unstained thin section impregnated with blue-dye-stained resin, Jurassic, Saudi Arabia, PPL, × 30.

Carbonate Sediments and Rocks Under the Microscope

A minor type of primary fabric-selective porosity is intraparticle porosity. This occurs most commonly in bioclasts where the chambers and other spaces occupied by soft parts or fluid during life have remained empty during burial, or have been only partially filled with sediment or cement. This is illustrated in **299** where the chambers of a micrite-walled foraminifer remain empty. The microfossil is embedded in carbonate mud sediment and the whole rock has been partially dolomitised.

Fenestrae, by definition, are pore-spaces larger than normal grain-supported spaces. They form through a combination of desiccation and entrapment of gas bubbles in the sediment. Fenestral porosity most commonly occurs in carbonate mud-rich sediments of tidal flats, often associated with the decaying organic matter of cyanobacterial mats. Fenestrae of this type, associated with stromatolites (p.100), are characteristically elongate parallel with the sediment lamination and are known as laminoid fenestrae. An example is illustrated in **300**. The impregnating blue-dye-stained resin is very pale coloured in this case.

299 Unstained thin section impregnated with blue-dye-stained resin, Jurassic, Saudi Arabia, PPL, × 55.

300 Unstained thin section impregnated with blue-dye-stained resin, Quaternary, Bahamas, PPL, × 42.

Porosity

301 and **302** show further examples of laminoid fenestrae, although these fenestrae are filled with sediment and cement. In **301** cement-filled fenestrae parallel with bedding are associated with subvertical fractures which are probably filled desiccation cracks. In **302** there are abundant fenestrae, the largest of which are laminoid. In this example the fenestrae are filled with a mixture of silt-sized sediment and cement. The sediment rests on the base of the fenestrae and was deposited after the enclosing carbonate mud. Such sediments filling cavities are known as internal sediments, and where the filling is partial, such that the top surface records the horizontal at the time of deposition, they are also known as geopetal infills (see also **246, 257, 258**).

301 Unstained thin section, Upper Jurassic, Provence, France, PPL, × 36.

302 Stained thin section, Lower Carboniferous, Derbyshire, England, PPL, × 22.

Secondary mouldic porosity, formed through the selective solution of aragonite grains, has already been illustrated for ooids (**14,15**) and molluscs (**57**) as well as in this section (**298**). In **303** an aragonite mollusc fragment with a crossed-lamellar structure (p.36) is undergoing solution, with the left-hand part particularly dissolved. Further exposure to meteoric water would lead to the complete solution of this fragment and the formation of a large mouldic pore, in addition to the substantial inter-particle porosity already present. Many mouldic pores are later infilled with cement. This has happened in **304**, where micritised bioclasts have been dissolved, leading to a large increase in porosity. The pores have subsequently been filled with cement, such that the sediment now has no visible porosity. The porosity is sometimes said to have been occluded by the precipitation of the cement.

303 Unstained thin section impregnated with blue-dye-stained resin, Quaternary, Pembrokeshire, Wales, PPL, × 24.

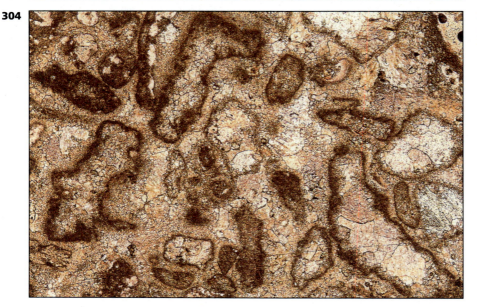

304 Stained thin section, Lower Jurassic, South Wales, PPL, × 35.

Porosity

Porosity may be increased at or soon after the time of deposition by the activities of boring organisms. **305** shows small borings on the surface of a single ooid grain, whereas **306** is an example of a boring that has cut into cemented rock, in this case an ooid packstone. In **306** the sediment must have been cemented at the time of boring because the ooids have been cut through (upper right of boring, for example), rather than having been pushed aside. The boring was later filled with a different sediment, mostly of fragmented bioclasts.

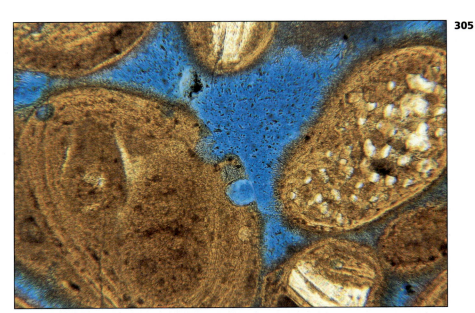

305 Unstained thin section, impregnated with blue-dye-stained resin, Quaternary, Kuwait, PPL, × 180.

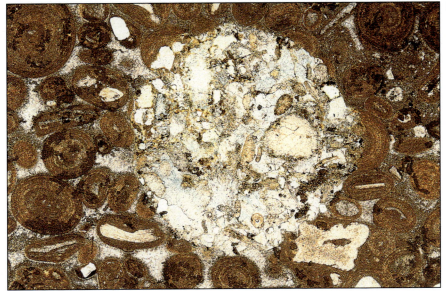

306 Stained thin section, Middle Jurassic, Gloucestershire, England, PPL, × 22.

307 shows fracture porosity. This is clearly not fabric-selective since it cuts across both carbonate mud and dolomite rhombs. **308** shows intercrystalline porosity in a dolomite. This type of porosity is regarded as secondary, although there is no way of knowing whether any of the porosity was inherited from the precursor limestone.

309 and **310** show solutional porosities which are fabric-selective. In **309** a calcite cement shows signs of solution in the form of small cavities within the crystals. The rounded outline of some of the crystals may also be solutional in origin. Small-scale porosity, in which pores are generally less than 1/16 mm in diameter, is called microporosity. In **310** it is dolomite crystals which have undergone partial solution. This is a form of dedolomitisation (p.147), a process often associated with Recent uplift and weathering. However, this is not the case here, since the sample comes from a core taken from a depth of several thousand feet. There is also some preserved primary interparticle porosity.

Non-fabric-selective solutional porosity can be classified as vuggy, channel-like or cavernous, according to the shape and size of pores. This type of porosity is often associated with uplift and karstification of limestone successions. Vuggy porosity can also be associated with hydrocarbon reservoirs where the agent of solution may be aggressive oilfield formation waters. **311** shows a vuggy porosity associated with patchy hydrocarbon residues (dark brown or black). Note that this section was not impregnated with blue-dye-stained resin and the pores therefore appear white in the photograph.

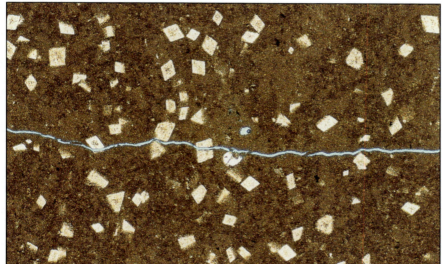

307 Unstained thin section impregnated with blue-dye-stained resin, Jurassic, Saudi Arabia, PPL, × 52.

308 Unstained thin section impregnated with blue-dye-stained resin, Devonian, Alberta, Canada, PPL, × 42.

Porosity

309 Unstained thin section impregnated with blue-dye-stained resin, Upper Cretaceous, Denmark, PPL, × 100.

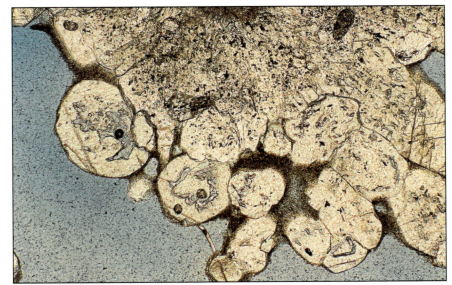

310 Unstained thin section impregnated with blue-dye-stained resin, Jurassic, Saudi Arabia, PPL, × 60.

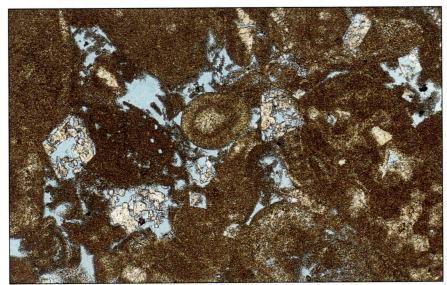

311 Stained thin section, Jurassic, Saudi Arabia, PPL, × 42.

LIMESTONE CLASSIFICATION

There are two widely used limestone classifications, those of Dunham (1962) and Folk (1959, 1962). The simplest descriptive classification is that of Dunham, where rocks are assigned names according to their depositional texture (312). This is primarily related to the energy of the depositional environment. The boundstone category, embracing sediments such as stromatolites and reef rocks which are bound into solid masses as they grow, has been subdivided, but these divisions are best recognised by large-scale hand-specimen and outcrop-sized features and are not considered here. The problems with the Dunham classification lie in the timing of introduction and origin of carbonate mud in packstones and the distinction of grain- and matrix-supported textures, as discussed by Tucker & Wright (1990). The classification is based on *depositional* texture and it is often difficult, if not impossible, to determine whether carbonate mud in a packstone was introduced at the time of deposition or subsequently infiltrated a primary grainstone.

To illustrate these problems two apparent packstones are illustrated in **313** and **314**. Each contains both sparry calcite cement and carbonate mud with rather more carbonate mud than cement. However, in **313** there is a distinct isopachous crust of cement which surrounds the bioclasts and which predates the introduction of the carbonate mud. Clearly this sediment was a grainstone when deposited. The origin of the mud in **314** is not so clear. It may have been deposited along with the grains, and the sediment would therefore be a genuine depositional packstone, or it may have infiltrated later and have been a grainstone as in **313**. A further possibility exists with some packstones. Depositional mud-supported textures (wackestones) can be converted to packstones by water loss from the mud during compaction, the result being a grain-supported texture. There is no 'magic' boundary between packstones and wackestones, and within one thin section it is not uncommon to see both textures.

312

Original components not organically bound together during deposition				Components organically bound during deposition
Contains carbonate mud			No carbonate mud	
Mud-supported		Grain-supported		
<10% allochems	>10% allochems			
MUDSTONE	WACKESTONE	PACKSTONE	GRAINSTONE	BOUNDSTONE

312 Classification of limestones according to Dunham (1962). Rock names are in capital letters.

Limestone Classification

Despite these problems, the Dunham classification remains the most popular simple classification of limestones. The textural name can be prefaced by the name of the principal grain type and sometimes by other textural information to give an unambiguous name that conveys in a few words the essence of a particular sediment. For example, the depositional textural classification of **313** and **314** would be bioclastic grainstone and poorly sorted oolitic peloidal packstone.

The full range of textures is not illustrated here, since they have been illustrated elsewhere in this atlas. In particular, the reader is referred to the following examples for good examples of the main textural types:

6 Well-sorted oolitic grainstone.
45 Bioclastic packstone.
60, 61, 66, 69 Bioclastic wackestones.
121, 125, 145, 155 Bioclastic grainstones.
29, 30 Peloidal grainstones.
32 Peloidal wackestone.
37 Intraclastic, bioclastic, oolitic grainstone.
235 Mudstone.

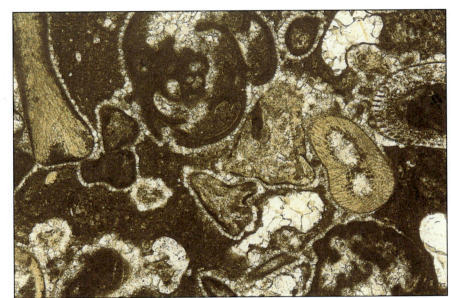

313 Unstained thin section, Lower Carboniferous, Lancashire, England, PPL, × 35.

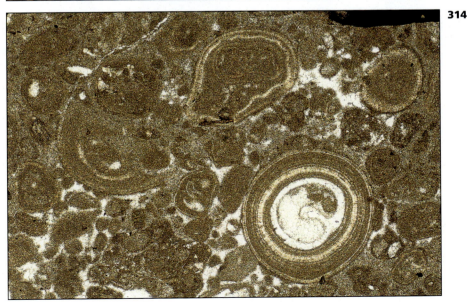

314 Unstained thin section, Upper Jurassic, Provence, France, PPL, × 45.

Carbonate Sediments and Rocks Under the Microscope

It must be borne in mind that rock names should not normally be assigned from a single field of view, but embrace the whole of the thin section under study.

The Folk classification of limestones is outlined in **315** and **316**. The basic limestone types are:

1. Those with the grains (allochems) set in a sparite cement.
2. Those with grains in a micrite matrix.
3. Micritic limestones lacking grains.

4. Organic limestones such as stromatolites and reef rocks equivalent to Dunham's boundstone.

With the allochemical rocks of categories 1 and 2, Folk derived a name based on a combination of part of the name of the dominant allochem (one of ooid, peloid, intraclast and bioclast) and whether the sediment was cemented by sparite or had a micrite matrix, hence oosparite, pelmicrite, etc. (**315**).

315

Volumetric allochem composition			>10% allochems		<10% allochems		
			Sparry calcite > Micrite	Micrite > Sparry calcite	1–10% allochems	<1% allochems	Undisturbed reef and bioherm rocks
	>25% Intraclasts		INTRASPARITE	INTRAMICRITE	Intraclasts INTRACLAST-BEARING MICRITE		
<25% Intraclasts	<25% Ooids	>25% Ooids	OOSPARITE	OOMICRITE	Ooids OOID-BEARING MICRITE	MICRITE, or if sparry patches present DISMICRITE	
	Volume ratio, bioclasts: peloids	> 3:1	BIOSPARITE	BIOMICRITE	Bioclasts FOSSILIFEROUS MICRITE		BIOLITHITE
		3:1 to 1:3	BIOPELSPARITE	BIOPELMICRITE			
		< 1:3	PELSPARITE	PELMICRITE	Peloids PELOID-BEARING MICRITE		

315 Classification of limestones based on the scheme of Folk (1959, 1962). Rock names are in capital letters.

Limestone Classification

Folk further refined the classification by introducing a textural element (**316**) dependent on the proportion of grains, micrite matrix and cement, and on the degree of rounding and sorting. As with the Dunham classification, the idea is to gain information about the energy levels in the depositional environment from the rock name. Some of the Folk names for limestones illustrated in this atlas are:

6 Sorted oosparite.
10 Poorly washed oosparite.
45 Packed biomicrite.
60, 66 Sparse biomicrite.
121 Sorted biosparite.
30 Unsorted pelsparite.
37 Unsorted intrasparite.
235 Fossiliferous micrite.
314 Poorly washed pelsparite.

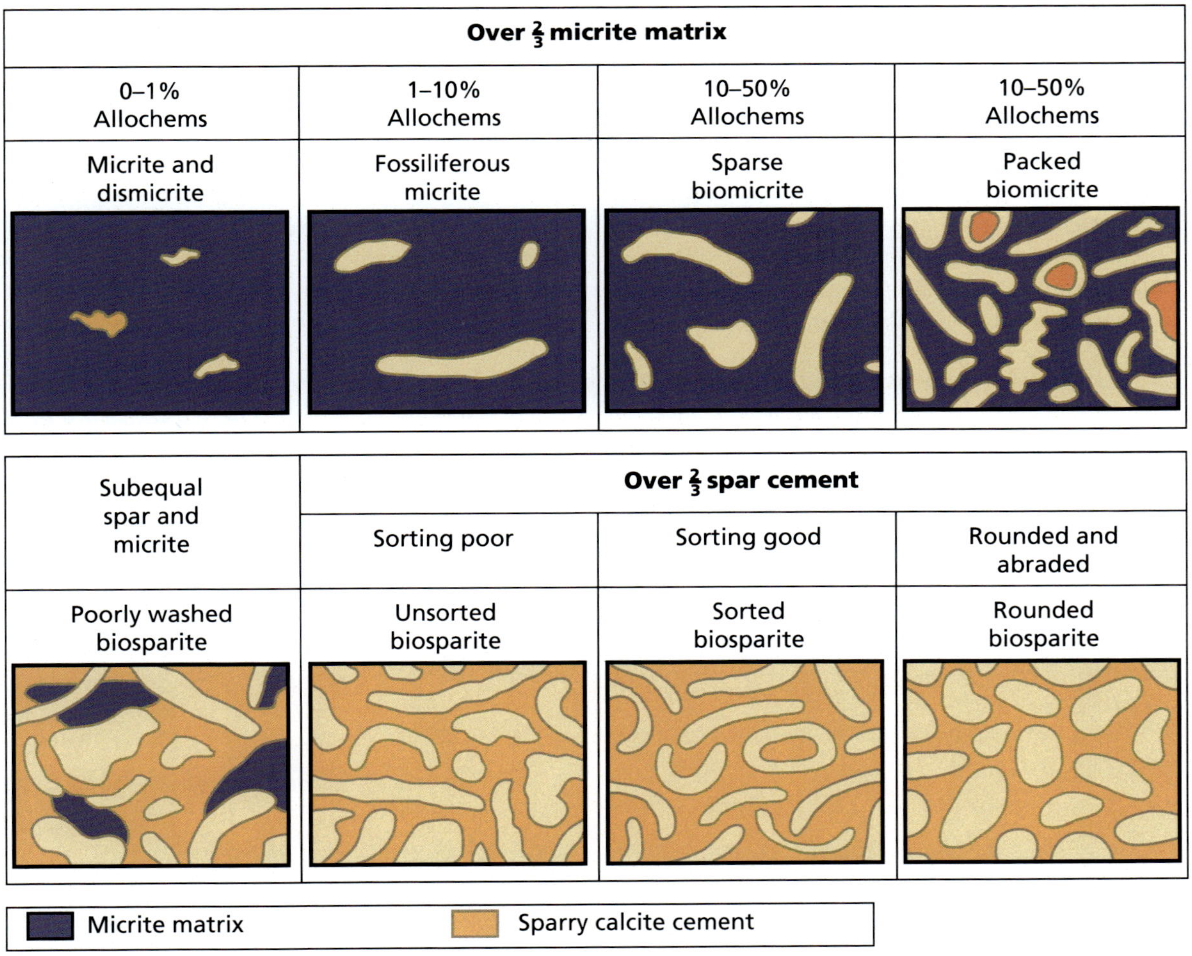

316 Subdivision of limestone types according to texture (after Folk, 1959).

CATHODOLUMINESCENCE

Some natural materials emit visible light when bombarded with an electron beam and this is the phenomenon of cathodoluminescence (CL). Carbonate minerals are particularly prone to luminescence, and since ordinary polished thin sections and relatively inexpensive equipment are needed, the technique has become a routine part of carbonate petrography. It is impurities within the carbonate minerals, rather than the major elements, which give rise to most of the visible luminescence. The most important ions affecting luminescence intensity in carbonates are Mn^{2+} and Fe^{2+}, with the manganese activating luminescence and the iron quenching it. Hence, variations in luminescence intensity usually reflect a variation in the ratio of Mn^{2+} to Fe^{2+} in a crystal. Such changes reflect variations in pore-water chemistry or precipitation mechanism. CL studies are a bridge between ordinary petrographic studies and micro-chemical analysis. CL does not reveal absolute concentrations of trace elements, but helps characterise generations of cement and other diagenetic minerals for further analysis. An introduction to cathodoluminescence and its use in carbonate sedimentology can be found in Miller (1988).

317 and **318** show what may be regarded as a typical cement sequence in a limestone cemented by calcite precipitated from meteoric water. The ordinary light view (**317**) shows a grainstone with a drusy mosaic cement of equant crystals. The very open texture suggests that cementation began early in diagenesis. This is the type of cement that would be interpreted as a product of meteoric phreatic diagenesis (p.104). With CL (**318**), two main generations of cement are visible. The first, and more abundant, is dark, and the second shows orange luminescence of moderate intensity. Separating these generations, there is a thin bright yellow zone, well seen just below the centre of the photograph. Dark/bright/dull luminescent zonation is known from many cements. The dark zone is more or less free of Mn^{2+} and Fe^{2+}; the bright zone contains the activator, Mn^{2+}, but not the quencher, Fe^{2+}, and the dull zone (or moderately luminescent cement in the sample illustrated) contains both Mn^{2+} and Fe^{2+}. One interpretation of this is that it reflects decreasing Eh, i.e. the solution changing from oxidising to reducing during increasing burial. The dark zone represents precipitation from oxidising waters containing neither Mn^{2+} nor Fe^{2+} in solution. As oxygen is used up, the conditions become briefly suboxic, when Mn^{2+} can exist in solution and is incorporated in the growing calcite crystal, but Fe^{2+} cannot. Thus a thin brightly luminescing zone is precipitated. When conditions become anoxic, both Mn^{2+} and Fe^{2+} are present in pore-fluids and are incorporated in the cement. The effect of both ions being present is to produce dull or moderately luminescing calcite depending on the exact proportions of activator to quencher.

Cathodoluminescence

317, **318** Unstained thin section, Lower Carboniferous, South Wales, × 45, **317** ordinary light, **318** CL.

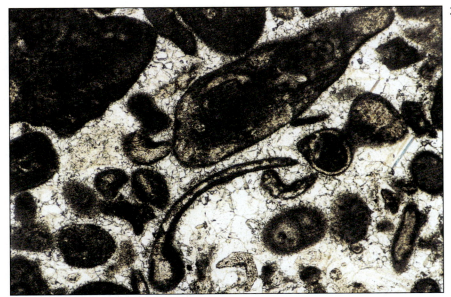

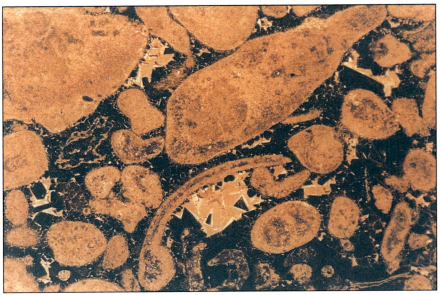

319 and **320** show the utility of cathodoluminescence in interpreting environments of cement precipitation. In the ordinary light view the cement fabric appears to be a drusy mosaic calcite very similar to that in **317**. However, with CL a different fabric is revealed. It can be seen that in the pores in the middle part of the photograph the dark cement has a distinct meniscus fabric, indicating precipitation in the vadose zone (p.104). The moderately luminescing orange-red cement fills the remaining pore-space. In this case there is no distinct bright zone at the contact between the two cement generations. This may be because initial cementation took place in the near-surface vadose zone and there was an hiatus before cementation resumed in the deeper subsurface. Thin veins are also visible in the CL view which are not evident in ordinary light.

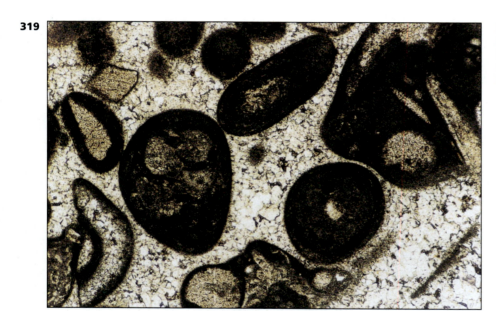

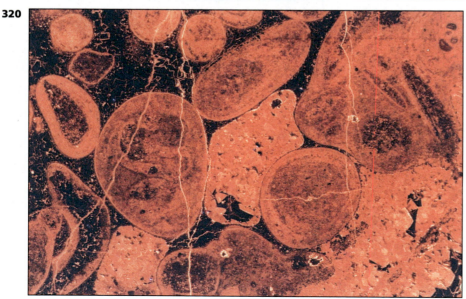

319, 320 Unstained thin section, Lower Carboniferous, South Wales, × 45, **319** ordinary light, **320** CL.

Cathodoluminescence

Cementation patterns revealed by CL can often be a lot more complicated than those in **318** and **320**. **321** shows a large shelter pore filled with a calcite cement exhibiting a drusy mosaic. The complexity of the chemical zonation in these crystals is revealed in the CL photograph (**322**). The rapidly alternating dark and moderately luminescing zones may result from precipitation from pore-waters with fluctuating chemistry, but may also be caused by disequilibrium precipitation where the trace element concentration is not that theoretically expected from a fluid of a particular composition.

321, **322** Unstained thin section, Lower Carboniferous, South Wales, ×45, **321** ordinary light, **322** CL.

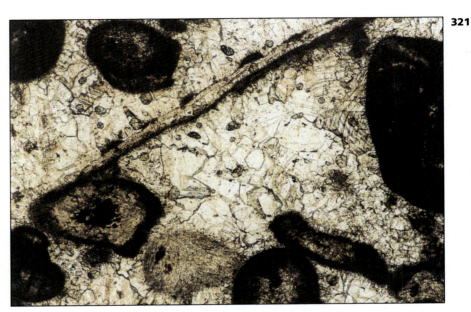

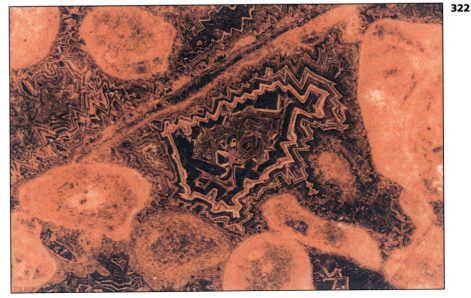

In the section on cementation, it was explained that echinoderms are often a preferred site for the precipitation of cement and that syntaxial overgrowth cements may develop at the expense of cements elsewhere in a rock (p.118). Because of this, syntaxial overgrowths may show a more complete picture of the cementation history of a sediment than the cement present on non-echinoderm substrates. **323** and **324** show a syntaxial overgrowth in a Carboniferous grainstone. Initial cements are dark in the CL view (**324**), although there is a hint of fine banding visible, particularly in the dark area just above the centre of the photograph. The lines running NW–SE are the calcite twinning and can be seen on both photographs. Note that the cement growth was faster, and cements better developed, on certain crystal faces. After the dark initial cement there is a dark cement with distinct bright yellow bands, seen only on the left side of the echinoderm, and finally a moderately luminescent cement which is clearly banded.

323, **324** Unstained thin section, Lower Carboniferous, South Wales, × 45, **323** ordinary light, **324** CL.

Cathodoluminescence

Characteristically, while calcites that luminesce show shades of yellow and orange, dolomites tend to be red, although there are many variations. **325** and **326** show a sediment that is a mixture of dolomite and calcite. The dolomite rhombs are evident in the ordinary light view (**325**), where they can be seen to be cloudy, although the centres of some, such as the large rhomb to the right of centre are rather less cloudy. In the CL view (**326**) the red-luminescing dolomite crystals are even more obvious and it can be seen that the corroded(?) cores of the crystals are much darker. Calcite between the dolomite crystals shows shades of yellow and orange luminescence.

325, **326** Unstained thin section, Lower Carboniferous, Derbyshire, England, × 45, **325** ordinary light, **326** CL.

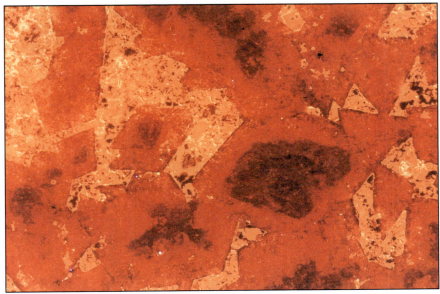

Carbonate Sediments and Rocks Under the Microscope

Dolomites can show complex zoned patterns although often not as spectacularly as in **328**. The ordinary light view of this rock (**327**) shows large dolomite crystals with clear centres and much cloudier rims (note that this is the reverse of the normal relationship, p.142) and detrital quartz (the clear crystals in the centre left and upper right of the photograph). In the CL view (**328**) the central parts of the crystals contain a lot of bright yellow zones. This colour is more characteristic of calcite than dolomite, and it may be that in these zones the luminescence activator (Mn^{2+}) is in the calcium site rather than the more normal magnesium site. Later zones are dominantly red. Note that the faces showing more rapid crystal growth appear to have changed during dolomite formation. CL often reveals complex patterns, the geological significance of which is difficult to determine.

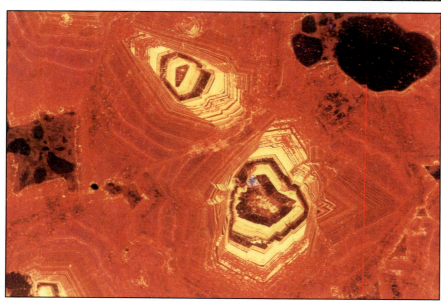

327, 328 Unstained thin section, Cambrian, Senegal, × 45, **327** ordinary light, **328** CL.

Cathodoluminescence

329 shows a mosaic of calcite crystals and dark iron-oxide-rich areas, some of which delineate rhomb shapes (for example, in the right-hand part of the photograph). These are dedolomites (p.147), and the shapes are much more evident in the CL view (**330**). The dedolomite calcite is mostly dark, although some dark red colouration suggests that there may be some dolomite remaining within the iron-oxide-stained areas. Calcite cements between the former dolomites show initial bright yellow and later dark luminescence similar to that of the dedolomite.

329, 330 Unstained thin section, Lower Carboniferous, Derbyshire, England, × 45, **329** ordinary light, **330** CL.

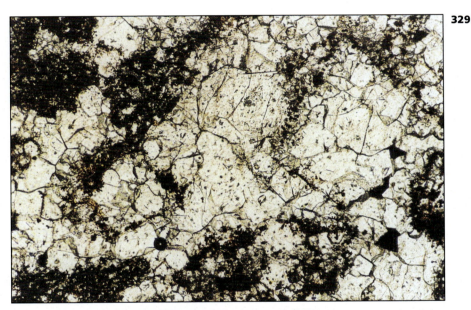

329

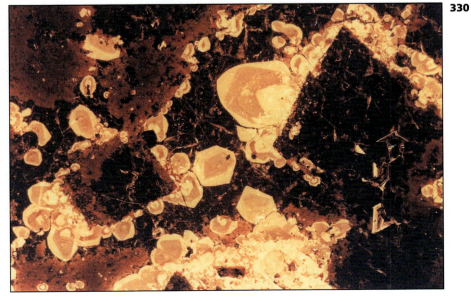

330

BIBLIOGRAPHY

Adams, A.E., MacKenzie, W.S. & Guilford, C. 1984. *Atlas of sedimentary rocks under the microscope*. Longmans, Harlow.

Bathurst, R.G.C. 1975. *Carbonate sediments and their diagenesis*. Elsevier, Amsterdam, 2nd edition.

Brasier, M.D. 1980. *Microfossils*. Allen & Unwin, London.

Choquette, P.W. & Pray, L.C. 1970. Geologic nomenclature and classification of porosity in sedimentary carbonates. *American Association of Petroleum Geologists, Bulletin*, **54**, 207–250.

Dickson, J.A.D. 1965. A modified staining technique for carbonates in thin section. *Nature*, **205**, 587.

Dunham, R.J. 1962. Classification of carbonate rocks according to depositional texture. In: Ham, W.E. (ed.) *Classification of carbonate rocks*. American Association of Petroleum Geologists, Memoir 1, 108–121.

Flügel, E. 1977. *Fossil algae*. Springer, Berlin.

Flügel, E. 1982. *Microfacies analysis of limestones*. Springer, Berlin.

Folk, R.L. 1959. Practical petrographic classification of limestones. *American Association of Petroleum Geologists, Bulletin*, **43**, 1–38.

Folk, R.L. 1962. Spectral subdivision of limestone types. In: Ham, W.E. (ed.) *Classification of carbonate rocks*. American Association of Petroleum Geologists, Memoir 1, 62–84.

Folk, R.L. 1965. Some aspects of recrystallizaion in ancient limestones. In: Pray, L.C. & Murray, R.C.

(eds) *Dolomitization and Limestone Diagenesis: A Symposium*. Society of Economic Palaeontologists and Mineralogists Special Publication, Vol. 13, 14–48.

Harwood, G.M. 1988. In: Tucker, M.E. (ed.) *Techniques in sedimentology*. Blackwells, Oxford, 174–190.

Horowitz, H.S. & Potter, P.E. 1971. *Introductory petrography of fossils*. Springer, Berlin.

Johnson, J.H. 1961. *Limestone-building algae and algal limestones*. Colorado School of Mines.

MacKenzie, W.S. & Adams, A.E. 1994. *A colour atlas of rocks and minerals under the microscope*. Manson, London.

Majewske, O.P. 1969. *Recognition of invertebrate fossil fragments in rocks and thin sections*. E. J. Brill, Leiden.

Miller, J. 1988. Cathodoluminescence microscopy. In: Tucker, M.E. (ed.) *Techniques in sedimentology*. Blackwells, Oxford, 174–190.

Peryt, T. (ed.) 1983. *Coated Grains*. Springer, Berlin.

Scholle, P.A. 1978. *A color illustrated guide to carbonate rock constituents, textures, cements and porosities*. American Association of Petroleum Geologists, Memoir 27.

Tucker, M.E. & Wright, V.P. 1990. *Carbonate sedimentology*. Blackwells, Oxford.

Wray, J.L. 1977. *Calcareous algae*. Elsevier, Amsterdam.

INDEX

All references are to page numbers.

acanthopores 62
agglutinated wall structure 67, 68
aggrading neomorphism 128, 130
aggregate grains 23, 27
algae 81–91
algal nodules 20
Alizarin Red S 7
alveolar septal fabric 102
alveolinids 70
ammonites 46
anhydrite 153, 154
ankerite 132
aptychi 46
aragonite 6, 35
archaediscids 70
arthropods 92–95

baroque dolomite 144–146
beachrock 108
belemnites 46, 47
beresellids 86
bioclasts 6, 32
bivalves 32–42
blastoids 76
blue-green algae 99
borings 156, 161
boundstone 164
brachiopods 48–53
   impunctate 48
   pseudopunctate 48, 50, 51
   punctate 48
   spines 51
bryozoans 62–66
   bifoliate 62, 64
   fenestrate 29, 65, 66

calcareous algae 81–91
calcispheres 86
calcitisation 128
calpionellids 97
cast 32, 41
cathodoluminescence 168–175
cementation 104
cements
   acicular aragonite 104
   blocky 114
   botryoidal 112
   burial 121

cements (*continued*)
   dripstone 108, 109
   equant 114
   fascicular optic 112
   isopachous 104, 106, 111, 114, 116
   marine 104, 106, 110, 114, 116
   meniscus 104, 106, 107
   meteoric 104, 109, 110
   micritic 104, 106
   peloidal 27, 104, 112
   phreatic 104, 111
   prismatic 104, 108
   pyrite 154
   radial fibrous 109, 111
   radiaxial 112
   silica 151
   syntaxial overgrowth 118–120
   syntaxial rim 118–120
   vadose 104, 106–109, 170
   zonation 171, 172
cephalopods 46
Chaetetes 61
chalcedony 152
charophytes 87
chert 149
Chlorophyta 81
Chrysophyta 81
coated grains 9–22
coccolithophorids 81
codiaceans 81, 84, 85
compaction 122–126
Corallinaceae 88
coralline algae 88
corals
   heterocorals 58
   rugose 54, 57, 58
   scleractinian 54, 56, 57
   tabulate 54, 57
cornstone 104
cortex 9
crinoids 76, 78–80
crossed lamellar wall structure 32, 36, 43
crystal silt 131
cyanobacteria 99, 100
Cyanophyta 81
cystoids 76

dasycladaceans 81–83
dedolomites 147, 148
deformation 127

degrading neomorphism 131
diagenesis 101
diagenetic environments 101
discocyclinids 72
disequilibrium precipitation 171
dolomites 132–146
    baroque 144–146
    cathode luminescence 173, 174
    equicrystalline 141
    ferroan 7, 8, 142, 145
    inequicrystalline 141
    mimetic 136
    mimicking 136
    non-planar 141
    planar 141
    saddle 144
    zoned 142, 174
dolomitisation 132–144
dripstone 108
drusy mosaic 110, 114, 117
Dunham classification 165, 167

echinoderms 76–80
echinoids 76–78
endothyraceans 70
etching 6, 7
evaporites 153, 154
extraclasts 23

fabric selective 132, 134, 156
fascicular optic calcite 112
Feigl's Solution 8, 35
fenestrae 100, 137, 156, 158, 159
ferroan calcite 7, 8
ferroan dolomite 7, 8, 143, 145
ferroan minerals 7
flat pebble conglomerate 137
fluorite 155
foliated wall structure 48
Folk classification 166, 167
foraminifera 67–75
Fusulinina 69, 70

gastropods 43–45
geopetal 131, 137, 159
Girvanella 99
glauconite 154
globigerinaceans 75
globigerinids 75
globotruncanids 75
goniatites 46
grainstone 164
grain-to-grain pressure solution 123

grapestones 23
green algae 81–87
gymnocodiaceans 88, 90
gyrogonites 87

Halimeda 84
hardground 110
homogeneous wall structure 32, 35, 37
hyaline wall structure 72

impregnation 8
impunctate 48
internal sediment 131
intraclasts 23, 27–29
isopachous 104

Koninckopora 82

laminoid fenestrae 137, 159
limestone classification 164
lithoclasts 23, 30, 31
luminescence 168
    activators 168
    quenchers 168

marine cements 104, 106, 110, 114, 116
meteoric cements 104, 109, 110
micrite envelope 101, 102
micritisation 101, 102
microbial structures 99, 100
Microcodium 102
micro-oncoid 9, 20
microspar 128, 130, 131
Miliolina 69, 70
mimetic dolomitisation 136
mimicking dolomitisation 136
molluscs 32–47
mould 40
mudstone 164

needle-fibre calcite 102
neomorphism 128–130
non-fabric selective 132, 136, 156
non-ferroan minerals 7
nucleus 9
nummulitids 79

oncoid 9, 20
oncolith 9
oogonia 87
ooids 9–18
oolith 9
oomouldic porosity 16

Index

orbitoids 72
orbitolinids 68
ossicles 78
ostracods 94, 95
oysters 33, 38

packstone 164
palaeoberesellids 86
pedogenesis 102
pelagic bivalves 42
pellets 23, 24
peloids 23–27
peneroplid 69
phreatic 101, 104
phylloid algae 85
pisoids 9, 19
pisolith 9
poikilotopic fabric 119, 121
porcelaneous wall structure 69
porosity 156–163
    burrows and borings 156, 161
    cavern 156
    channel 156
    fabric selective 156
    fenestral 131, 156, 158, 159
    fracture 156, 162
    intercrystal 142, 156, 162
    intergranular 156, 157
    interparticle 156, 157
    intragranular 153, 156, 158
    intraparticle 156, 158
    mouldic 16, 40, 156, 157
    non-fabric selective 156
    oomouldic 16
    primary 156
    secondary 156
    vuggy 156, 163
potassium ferricyanide 7, 8
pressure solution 123
prismatic wall structure 32–35, 48, 52
pseudo-pleochroism 128
pseudopunctate 48, 50
pseudospar 130
pseudo-uniaxial cross 10, 51
punctate 48
pyrite 154

quartz 10, 22, 31, 98, 104, 149-152, 174

radiaxial calcite 112
radiolarians 98
recrystallisation 128
red algae 81, 88-91

Renalcis 99
rhizocretions 102
rhodoids 88
rhodolith 88
Rhodophyta 81
rotaliaceans 73
Rotaliina 72, 74
rudists 39

Saccaminopsis 87
saddle dolomite 141
serpulids 96
shell structures
    bivalve 32–42
    brachiopod 48–53
    cephalopod 46, 47
    crossed-lamellar 32, 36, 43
    foliated 32, 33, 38
    gastropod 43–45
    homogeneous 32, 35, 37
    prismatic 32–35, 48, 52
silicification 149, 150
Solenoporaceae 88, 90
solenoporoid algae 90
spalling 125
sparite 6, 7
speleothem 108
spherulites 12
spines
    brachiopod 51
    echinoid 76–78
spicules 60, 66, 98
Spirorbis 96
sponges 60, 61
staining 6–8
stromatolites 100
stromatoporoids 59
stylolites 123, 134
superficial ooids 14
syntaxial overgrowth cements 118
syntaxial rim cements 118

taleolae 50
Textulariina 67
Tentaculites 97
tintinnids 97
trilobites 92, 93
twinning 127

undulose extinction 41, 112, 126, 144–146

vadose 101, 104
veins 126

Carbonate Sediments and Rocks Under the Microscope

vermiforms   96

wackestone   164

wall structures
   agglutinated   67, 68
   crossed lamellar   32, 36, 43
   foliated   32, 33, 38
   homogeneous   32, 35, 37
   hyaline   72
   impunctate   48
   porcelaneous   69

wall structures (*continued*)
   prismatic   33–35, 48, 52
   pseudopunctate   48, 50
   punctate   48
worm tubes   96

zonation
   cements   168–172
   dolomites   142, 174
zooecia   62